ADVANCES IN GLOBAL HIGH-TECHNOLOGY MANAGEMENT

Volume 1 • 1992

HUMAN RESOURCE STRATEGY IN HIGH TECHNOLOGY

ADVANCES IN GLOBAL HIGH-TECHNOLOGY MANAGEMENT

A Research Annual

HUMAN RESOURCE STRATEGY IN HIGH TECHNOLOGY

Editors: LUIS R. GOMEZ-MEJIA
Arizona State University

MICHAEL W. LAWLESS
University of Colorado at Boulder

VOLUME 1 • 1992

JAI PRESS INC.

Greenwich, Connecticut *London, England*

55 Old Post Road, No. 2
Greenwich, Connecticut 06836

JAI PRESS LTD.
118 Pentonville Road
London N1 9JN
England

ISBN: 1-55938-489-1

Manufactured in the United States of America

CONTENTS

INTRODUCTION

This volume focuses on human resource management systems in high technology firms. It is divided into three sections. The first one is concerned with macro linkages between human resource strategy and organizational strategy. Three papers are included in this section. The first one, by Martell, Carroll, and Gupta, reviews the role of human resource management (HRM) in strategy implementation, with a focus on innovation strategies. It also develops an empirical profile of HRM practices, based on a sample of 115 companies, that are associated with exceptional firm performance in innovative business environments. The second paper by Mohrman, Mohrman, and Cohen, examines human resource strategies that enhance lateral integration in a high technology setting. This calls for HRM practices that engender more reliance on self-control and deemphasize hierarchical relationships. The last paper in this section, by Masters, Tokesky, Brown, Atkin, and Schoenfeld, examines compensation strategies in the high technology industry. These authors discuss the role of pay-for-performance systems, various compensation techniques, pay secrecy, and employee involvement programs as part of a broader HRM strategy for high-technology organizations.

The second section of this volume is concerned with staffing issues in high technology firms, and consists of five papers. The first chapter, by Coombs and Rosse, examines current state of research into the recruitment and hiring processes utilized by high-technology firms. It identifies key issues that should be addressed in future research in this area. The second paper by Duchon, Ashmos, and Dunegan, develops an empirical profile of "innovators" and "noninnovators" (based on a sample of 239 scientists, engineers, and managers) in terms of demographic, psychological, behavioral, and role factors. The third paper, by Drory, focuses on intentions to quit among high technology professionals. He presents data from 200 organizations suggesting that a job design approach should be used to reduce turnover among professionals in the high technology industry given that the main predictor of withdrawal cognitions is intrinsic work characteristics. The next two papers are concerned with promotional practices in the high technology industry. Katz, Tushman, and Allen review the literature on dual career ladders and conclude that we still know little about how technical professionals make sense of these structures. Based on data collected from a sample of 345 technical staff members, Katz, Tushman, and Allen argue that organizational context affects the dynamics of a dual career ladder system and that these contextual factors should be incorporated into future research on this topic. Page, Stephens, and Tripoli examine successive changes in the psychological contracts between engineers and their employers at different stages of the entrepreneurial career path. Their analysis is based on a review of empirical findings from the careers and high technology/innovation literature and data from two in-depth case studies.

The third and last section of this volume focuses on training and development issues in a high technology environment. It includes three papers. The first one, by Bentley, argues that traditional training approaches are hopelessly inadequate or even dysfunctional in high-technology firms because the future is seldom similar to the past; therefore, one should not attempt to design learning experiences to deal with problems specified apriori. He provides a list of suggestions as to how high-technology firms can prepare employees to be successful in "permanent white water." The second paper by Kleiman and Foley argues that training programs in a high-technology setting would be more effective if one can first isolate task-based performance deficiencies that are correctable and then device specific training programs to address them. They illustrate how the appropriate content of performance based training programs can be identified based on a case study of a nuclear power division of a large public utility. The last paper in this section, by Liden and Adams, focuses on structural employment problems created as a result of an aging workforce in a world where technological change has become the norm. These authors present a theoretical model that deals with how the performance of older employees is affected by technological change and job redesign, how they respond to training, and the role that supervisors play in their adaptation process to technological change.

Luis R. Gomez-Mejia
Michael W. Lawless
Editors

PART I

HUMAN RESOURCE STRATEGY

WHAT EXECUTIVE HUMAN RESOURCE MANAGEMENT PRACTICES ARE MOST EFFECTIVE WHEN INNOVATIVENESS REQUIREMENTS ARE HIGH?

Kathryn Martell, Stephen J. Carroll, and Anil K. Gupta

INTRODUCTION

In today's increasingly competitive marketplace, innovation is frequently recommended as a solution for our most pressing business and economic problems (Drucker, 1985; Foster, 1986; Kanter, 1983). Scholars and business leaders encourage U.S. companies to become more consciously concerned with product and process innovation and often point to the Japanese as models for this orientation (Imai, 1986). Of course, smaller hi-tech companies are typically oriented this way. Even in traditional industries such as steel or autos, a company's success—not to mention survival—often is dependent on creating better products faster than the competition. For many companies, innovation is the key to meeting this challenge.

Human Resource Strategy in High Technology.
Advances in Global High-Technology Management, Volume 1, pages 3-30.

ISBN: 1-55938-489-1

Companies pursuing innovative strategies require special skills and talents from their employees, particularly from managers (Evans, Doze, & Laurent, 1989; Martin, 1984; Schuler & MacMillan, 1984). In addition to the technical expertise and leadership skills typically required of senior management, managers in innovative companies are often required to be creative, take risks and work in dynamic, uncertain environments. In short, innovative strategies require innovative managers (Souder, 1987), and it appears unlikely that a company could pursue the former without the latter. Academics and practitioners alike are drawing the conclusion that the "best laid plans" regarding innovation or other fundamental strategy decisions quickly become unworkable if requisite managerial talent is lacking.

How can innovative companies find the type of managerial talent they need to be successful? A number of theorists argue that a company's Human Resource Management (HRM) system can be used to attract, develop and motivate senior executives that are well-suited to specific strategic pursuits (Dyer, 1984b; Golden & Ramanujam, 1985; Schuler & Jackson, 1987b). According to this logic, a company's ability to tailor HRM activities—staffing, training, performance appraisal, and compensation—to "fit" its strategy helps determine whether they are able to successfully implement their business strategies. This proactive approach to HRM, which is often referred to as Strategic Human Resource Management, posits a linkage between strategy, HRM, and performance (Devanna, 1984; Dyer, 1984a, 1984b) and bestows on HRM the strategic importance previously reserved for financial and technological resources (Golden & Ramanujam, 1984; Tichy, Fombrun, & Devanna, 1984). According to this framework, strategic choice results in demand for certain behaviors and talents. In turn, HRM policies have the potential to modify and encourage workforce behavior (Baird, Meshoulam, & Give, 1983; Carroll, 1988; Fombrun & Tichy, 1983). Changes in direction may require changes in HRM systems. When the gap between needed and actual behavior is closed, strategy is effectively supported and performance is enhanced. Thus, policymakers need to identify behaviors and skills that correspond with their business strategies, and design HRM policies to elicit, build and maintain an appropriately skilled and commited workforce.

Much of the writing on the strategic aspects of HRM is consistent with Galbraith and Nathanson's (1978) strategy implementation model that states that in order for an organization's strategy to make an economic difference, strategy must be matched with an appropriate "organization form"—a combination of structure and process that includes all of the generic HRM functions. According to Galbraith and Nathanson, a change in strategy (defined in this model as a decision to develop new products or pursue new markets) introduces additional diversity into the corporate environment, which the organization must respond to by making changes in its various systems. If an organization resists the changes necessitated by the additional diversity and uncertainty, its financial performance will decline.

This model provides theoretical support for some of the major tenets that underlie strategic HRM models, namely:

1. HRM plays an important role in implementing strategy.
2. Ideally, strategic choice is a driving force determining HRM policies.
3. An organization's HRM practices will vary according to its strategic orientation.
4. Effectively matching HRM with strategy should result in an enhanced "bottom line" for the organization.

Furthermore, the model implicitly supports a "contingency theory" of HRM, hypothesizing that no one set of human resource policies is right in all strategic settings. HRM policies that contribute to the implementation of one type of strategy may prove ineffectual when matched with a different strategy. Thus, while firms pursuing innovative strategies may adopt different HRM policies, Galbraith and Nathanson's framework suggests that certain patterns of HRM decisions should emerge as maximally effective in innovative strategy settings.

A number of theorists have examined the role of HRM in implementing strategy (Carroll, 1987; Devanna, Fombrun, & Tichy, 1984; Dyer, 1984b; Gupta & Govindarajan, 1986; Labelle, 1983; Miles & Snow, 1984), and several of these studies have focused on innovation strategies (Lorange & Murphy, 1984; Martin, 1984; Schuler & Jackson, 1987a). Some investigators (Cook & Ferris, 1986; Kelleher & Cotter, 1982; Kerr, 1982; Perry, 1984; Smith, 1982; Tomasko, 1982) have linked various HRM practices to stages of the product life cycle. Others have linked very specific HRM practices such as staffing (Olian & Rynes, 1984) and compensation (Carroll, 1987, 1988; Milkovich, 1988; Stonich, 1981) to particular strategies. These writings—most of which are prescriptive in nature—focus on the linkage between strategy and HRM, and are consistent with a contingency theory of HRM which suggests that the effectiveness of HRM is dependent, in part, on the organization's strategic environment.

Other research has examined a second characteristic of a strategic approach to HRM—HRM's effect on organizational performance. Peters and Waterman (1982) are among those who have provided some anecdotal evidence that one characteristic of excellent U.S. companies is their high-quality HRM programs. Misa and Stein (1983) also provided anecdotal evidence that top performing companies demonstrated a more strategic approach to HRM than their lesser-performing counterparts. In a more empirical vein, Cook and Ferris (1986) found a relationship between performance and effective HRM in declining industries. Studies of specific HRM activities such as Dimick and Murray's (1978) study of staffing sophistication show variations in practices associated with organizational performance. Schuster's (1984) study of high tech companies concluded that higher-performing companies tailored compensation and performance evaluation systems to match strategic goals. Frohman (1984) also has evidence that certain types of HRM systems can effectively be used to implement new strategic directions of the company. This body of research suggests that HRM can have a noticeable effect on the company's "bottom line."

There are some shortcomings that limit the applicability of much of the strategic HRM research, however. First, despite the growing emphasis on linking HRM, strategy, and performance, little of the empirical research to date incorporates all three variables. Thus, much of the research does not provide a basis for examining the legitimacy of the HRM contingency theories which have grown in importance in the literature and in top management thinking. Including measures of organizational performance in studies previously focusing on the HRM/strategy relationship would add an important new dimension to this research, and could provide an empirical foundation for a normative model of HRM.

Second, while the prescriptive writings which dominate this subject area, contribute to theory development, and help set the stage for empirical research, many of the recommendations are rather vague and provide little specific guidance to managers on how to effectively design HRM systems to fit their companies' requirements. Also, some prescriptive models assume a state of HRM development well beyond current practice in even the best companies.

Finally, much of the empirical studies have utilized small samples with the resulting inability to utilize acceptable statistical methodologies (Lorange & Murphy, 1984). A more recent study by Schuler and Jackson (1987b) is a noteworthy exception in its use of a large sample and rigorous quantitative methodology. Other problems with the empirical research include simplistic operationalization of key concepts (particularly strategy) and various scaling difficulties.

PRESENT STUDY

This study focused specifically on the top executive teams in strategic business units (typically divisions or departments, with an average size of 4,792 employees) of major Fortune 500 companies in the United States. It examined the Executive Human Resource Management (EHRM) practices of companies in the four major HRM program areas: staffing, training, performance appraisal and compensation. EHRM practice variations were associated with a global organization performance measure within various innovative orientations of companies as indicated by their emphasis on new product development. The ultimate research goal was to identify patterns (if any) of EHRM practices associated with strong organizational performance in innovative business environments. We believe this is the most comprehensive empirical study of strategic Executive Human Resource Management practices using accepted measures of innovation in the literature to date.

Research Questions and Propositions

In order to better understand the linkage between innovation, EHRM, and performance, we addressed two issues in our research:

A. Are certain EHRM policies universally effective, regardless of organizational context? Is there a set of EHRM policies that due to their superior results in improving performance should be practiced by *all* companies, whether or not they pursue innovative strategies?
B. Are there EHRM practices that effectively support innovative strategies, but are ineffective or counter-productive in non-innovative companies? Is the effectiveness of EHRM policies dependent, in large part, on the strategic environment in which they are enacted?

These questions, while exploratory in nature, have their basis in emerging theories of HRM that posit a link between HRM and performance, and Galbraith and Nathanson's (1978) implementation theories that conclude HRM systems must "fit" with strategy. Our *a priori* expectations were expressed as research propositions, namely:

Proposition 1. Some individual EHRM practices are likely to have a discernable effect on organizational performance. Identifiable patterns of EHRM practices will be positively associated with high organizational performance in the sample of businesses incorporated in this study.

Proposition 2. A company's innovation strategy—as reflected by new product development—will have a significant moderating effect on the relationship between EHRM practices and performance. More specifically, in this study, identifiable patterns of EHRM practices will correlate with high performance for companies with relatively strong innovation strategy orientations.

Thus, in this study we expected to find both universal and contingent relationships. In terms of the latter, while we had certain expectations about the relationships that could be expected to be effective in innovative business environments (for example, we expected incentive compensation to be important in innovative companies), in the early stages of theoretical development characteristic of this research area it is probably premature to state these expectations in terms of specific hypotheses. Thus, rather than put undue emphasis on specific EHRM practice findings, our primary focus was to examine the linkage between EHRM and performance, and to test the validity of the more general Galbraith and Nathanson strategy propositions arguing for a contingency approach to EHRM. Individual findings, while potentially useful in the practice of EHRM, are not essential to testing theory in this area and need further replication before forming the basis for EHRM prescriptions.

METHODOLOGY

Sample

The sampling strategy used in this study closely adhered to the characteristics of a random sampling procedure that is required for the statistical techniques employed. The unit of analysis is the business unit (typically a division), and questionnaires were mailed to General Managers in a randomly selected group of Fortune 500 companies (as identified by Fortune magazine); business units and their general managers were listed in the *Book of Corporate Affiliations* (1988). General Managers were solicited for participation rather than HRM managers because they appeared to be in the best position to provide information on all critical variables: organizational performance and innovation strategy as well as HRM policies geared to their top executives. Subsequent analysis based on data collected supported this decision: for all of the EHRM practices included in our study, General Managers were the key decisionmakers. Thus, at least where executives are concerned, General Managers appear to have a great deal of autonomy in the HRM policy arena.

Twenty-six percent of the 450 questionnaires sent out were returned for a total of 115 useable responses. The study sample—which was heterogeneous with respect to industry and geographical location—closely approximated the characteristics of the random sample, thereby reducing the possibility of nonresponse bias. A description of the study sample is provided in Table 1.

Table 1. A Description of Sample Business Units
Size of Participating Organizations

Mean: 4,792 Employees
Standard Deviation:850
Minimum:4* Maximum 50,000

Industry Classification	*Percent*
Food and kindred products	6
Textile mill products	2
Apparel and fabric products	3
Paper and allied products	8
Printing and publishing	1
Chemicals and allied products	13
Petroleum refining	1
Rubber and miscellaneous plastic products	2
Leather and leather products	1
Stone, clay, and glass products	4
Primary metals	2
Fabricated metal products	2
Nonelectrical machinery	12
Electrical machinery and supplies	12
Transportation equipment	6
Scientific instruments, medical and optical	10
Miscellaneous manufacturing	2
Wholesale trade	7
Geographic Location	
Northeast	37
Southeast	10
Midwest/Central	37
Northwest	5
Southwest	11

Note: *Only 4 participants have less than 100 employees.

MEASURES

Innovation Strategy

Our innovation strategy measure focused on a dimension of innovativeness that is emphasized in much of the strategy literature: new product development (Dess & Davis, 1984; Miles & Snow, 1978; Porter, 1980). This conceptualization is also consistent with how Galbraith and Nathanson defined strategy in the implementation model that provides the theoretical underpinnings for this research. Managers were asked to indicate the percentage of the previous three years' sales that were derived from new products and from existing products. Thus, innovation was represented as a continuum, with businesses totally devoted to existing products at one end of the spectrum and companies with a strong new product orientation (up to 100% of sales) at the other. By focusing on previous years' sales, this measure reflects the organization's actual—not intended—strategy which is believed to more accurately reflect the true strategic orientation of the organization, thereby strengthening the construct validity of the innovation strategy measures. The range of responses was 0 to 100%, with a mean of 32.8% and standard deviation of 21.9%.

As a check on the validity of this measure, managers were asked to characterize the competitive strategy of their business unit using strategy descriptions that are prominent in strategy research. The innovative (new product) measure correlated significantly with a measure of competitive strategy based on Porter's (1980) generic definitions, where a high score represented a Differentiation posture typically emphasizing innovation and a low score represented a Cost Leadership posture ($r = .37; p < .001$). Finally, new product development typically requires research and development; we found our innovation measure to be significantly correlated with the importance accorded to the R&D function ($r = .37; p < .001$); likewise, the importance of new product development as a measure of business unit performance was positively correlated with our innovation measure ($r = .38; p < .001$).

Organizational Performance

The bulk of information published on the financial status of business organizations is only available at the corporate level. Given the highly sensitive nature of information pertaining to the financial position of individual business units (e.g., balance sheet items such as "Return on Assets" or "Net Profitability"), such data are considered by experienced researchers to be virtually impossible to obtain at the business unit level (Gupta, 1984; Lawrence & Lorsch, 1967). Thus, researchers are compelled to forgo financial measures when analyzing business units' performance and employ measures based on self-reported ratings. It should be noted that performance measures based on self-reported data have been found to be reliable if anonymity is guaranteed and participants understand that the objective of data collection is scientific, not evaluative (Heneman, 1974)—conditions that were fully met in this study.

There is increasing recognition in the literature that business performance is a complex and multidimensional phenomenon that incorporates diverse achievements (Ford & Schellenberg, 1982; Lenz, 1981) and we designed our performance measure to reflect this broader conceptualization. Adopting a methodology employed by Gupta and Govindarajan (1984), business unit performance was measured by a comparison between actual performance and *a priori* expectations along 12 different financial and nonfinancial criteria. Respondents were asked to indicate on a Likert-type scale: (1) the degree of importance that superiors attach to, and (2) their perception of performance relative to expectations on each of 12 different dimensions of organizational performance. Using the data on importance as criterion weights, a weighted average measure representing overall effectiveness was created for each business unit. The range on this measure was 2.4 to 6.5, with a mean of 4.4 and standard deviation of .9, indicating that as a group respondents did not appear to have inflated their reported performance. Gupta and Taylor (1989) performed a validity check of this particular performance scale by comparing self-reported data on financial measures with published data (where available). The self-reported data on Return on Assets was significantly correlated with the published data ($p < .01$). Thus, self-reported performance measures can have strong construct validity while significantly increasing the pool of business units as potential research participants.

Executive Human Resource Management Practices

The focus of this study is HRM programs geared toward executives, and we coined a term to reflect this distinction: Executive Human Resource Management (EHRM). The questionnaire gathered data on diverse EHRM activities that are cited in the literature as "core" fundamentals of personnel administration in business organizations (Milkovich & Boudreau, 1988) that appear to be particularly germane to the treatment of executives. Data was gathered on 19 items pertaining to decisions in each of the four generic HRM policy areas: staffing, training and development, performance evaluation, and compensation. The specific measures used included all of the following (more detail is provided in the Appendix):

1. *Staffing*
   - Source of managers
   - Importance of specific managerial skills
   - Importance of managers' innovative personality characteristics
   - Desirability of tenure with company
2. *Training and Development*
   - Participation in training activities (formal, informal, conferences)

3. *Performance Evaluation*
   - Emphasis on outcomes
   - Use of objective measures
   - Use of subjective measures
   - Frequency of formal appraisals
   - Frequency of informal appraisals
4. *Compensation*
   - Market competitiveness of compensation
   - Internal equity of salaries
   - Bonuses: Are bonuses paid?
     Frequency of bonuses
     Bonus amount
     Emphasis on individual incentives
     Subjectivity of bonus calculation

Some theorists suggest that HRM activities may vary across functional areas reflecting the strategic importance of individual functions (Dyer, 1984a); therefore, EHRM data was gathered separately for each of the three senior-most executives in charge of Research and Development (R&D), Marketing and Manufacturing. Subsequent analysis revealed that there was a high degree of uniformity in EHRM within senior management ranks, however, so for each company data for individual executives were combined into a single overall measure covering the "top management team" for each EHRM practice.

Statistical Analysis

The research propositions in this study focus on the association among variables:

- individual EHRM practices and organizational performance; and
- the interaction of EHRM and innovation strategy on organizational performance.

Regression analysis was used to evaluate both sets of proposition. To examine the association of EHRM and performance, multiple regressions were run for each of the 19 EHRM policies, controlling for strategy effects. The analysis was then expanded to incorporate the interaction effects of innovation strategy and EHRM on performance. This analysis, which examines EHRM's contingency effects, was performed through the use of moderated regression analysis (MRA). MRA involved regressing a moderating variable or interaction term (the crossproduct of innovation strategy and the EHRM practice) and the EHRM and strategy variables against organizational performance. A statistically significant regression coefficient for the interaction term indicates that there is a significant improvement in the variance explained by adding the moderator variable. These results are interpreted to mean that innovation strategy moderates the EHRM/performance relationship (i.e., the EHRM/performance relationship in the more innovative companies in our sample is different than the EHRM/performance relationship in less-innovative company environments).

THE UNIVERSAL RELATIONSHIP BETWEEN EHRM PRACTICES AND PERFORMANCE

Results

Nineteen separate regressions were run to examine the linkages between organizational performance and diverse EHRM practices: staffing, compensation, training, and performance appraisal. Of the 19, seven (37%) statistically significant relationships at the .05 level were identified through regression analysis, the majority of which are in the performance evaluation practices area. Table 2 summarizes these findings and the implications of the results are discussed below.

Staffing

Our staffing variables focused on selection criteria used in recruiting top managers (tenure, management skills, and personal characteristics) and recruitment method (whether managers were recruited from inside or outside of the firm). Of the four relationships analyzed, two were found to be statistically significant. The first of these was the extent to which diverse personal characteristics and manager's tenure with the organization were emphasized in recruiting managers. The second was tenure in the organization. Both were found to be positively associated with performance.

The personal characteristics measures considers eight diverse personality attributes generally recognized as relevant to managerial performance, including:

- creativity
- charisma
- adaptability
- mental toughness
- stress tolerance
- initiative
- willingness to take risks
- tolerance for ambiguity

Table 2. Results of Regression Analysis
Impact of EHRM Practices on Organizational Performance
(Test of Proposition 1)

	Standardized Regression Coefficients			
Independent EHRM Variable (X^1)	*Innovation*	*EHRM* (X^1)	R^2	*F-Value*
Staffing				
Source of New Hires	.001	.010	.000	.004 (2,83)
Imp. of Managerial Skills	.032	.167	.029	1.303 (2.88)
Imp. of Innovative Characteristics	.032	.211*	.046	2.195 (2,91)
Managers' Tenure	.023	.188*	.034	1,510 (2.85)
Compensation				
Are Managers Paid a Bonus?	.014	.152	.024	1.141 (2,94)
Bonus Based on Indiv. Performance	.094	.036	.009	.383 (2,81)
Potential Bonus Amount	−.002	.070	.005	.199 (2,82)
Use Formulas to Calculate Bonuses	.029	.021	.002	.064 (2,81)
Frequency of Bonus Payments	−.020	.061	.004	.167 (2,80)
Internal Equity of Salaries	.021	.159	.025	1.125 (2,88)
Competitiveness of Compensation	−.025	.276**	.075	3.449* (2,85)
Performance Appraisal				
Frequency of Formal Appraisals	.028	.067	.006	.273 (,289)
Frequency of Informal Appraisals	.053	.200*	.042	1.912 (2,87)
Focus on Outcomes	.041	−.060	.005	.208 (2,85)
Use of Objective Criteria	.008	.220*	.049	2.259 (2,88)
Use of Subjective Criteria	.032	.388***	.153	7.842*** (2,87)
Utilization of Evaluation Results	.002	.338***	.114	6.522*** (2,101)
Training				
Participation in Training and Development Activities	.030	.149	.024	1.082 (2,87)

Notes: * $p < .05$.
** $p < .01$.
*** $p < .001$.

These characteristics were chosen in part because of their apparent relevance to managing within innovative environments, and our *a priori* expectations were that only top-performing *innovative* organizations would emphasize these traits in choosing managers. This expectation was based on the belief that in addition to considering technical skills and relevant professional and educational experience, recruiters must decide what personality characteristics are most desirable for the position. We know from previous research that managers are different from a personality perspective, and that certain characteristics seem to be beneficial in certain strategic environments, while unhelpful in others (Gupta & Govindarajan, 1984; Szilagyi & Schweiger, 1984). In recent years, for example, managers such as Jerry Junkins (Texas Instruments), Jack Murphy (Dresser Industries), Walter Williams (Bethlehem Steel) and George Schaefer (Caterpillar Inc.) have gained a reputation as "turnaround" managers who have successfully rescued an organization from the brink of disaster. Placing such an individual in a management job in a stable organization may neither be the best use of such talent, nor in the organization's best interest. Recognizing that implementing strategies successfully (such as innovation strategies) require specific managerial leadership skills, and that these skills are in part personality-driven, leads to the conclusion that managers should be matched with strategic tasks, and this "match" should include relevant personality traits. Research studies indicate that such a match enhances an organization's performance (Gupta & Govindarajan, 1984).

Our expectation that we would find a contingency effect when examining the performance/characteristics relationship was not supported, however. Instead we found that the importance of several innovative personality characteristics were positively associated with strong performance in *all* organizations ($p = .05$). Thus, top performing organizations in our sample placed a high premium on creativity and the ability to deal with uncertainty when putting together their top management team, whether or not they were pursuing a new product innovation strategy. This finding tends to reinforce the emphasis on innovation in the business press that increasingly links an organization's ability to innovate with its competitiveness, or even survival. While business researchers tend to associate product innovation more strongly with certain strategies (e.g., Porter's Differentiation strategy, or Miles and Snow's Prospector strategy), today's increasingly competitive international market place may compel *all* organizations to be innovative, regardless of their strategic goals. Our finding of a universal relationship between the importance given to innovative personal characteristics (such as stress tolerance and willingness to take risks) and organizational performance suggests that all firms benefit from the leadership of aggressive, creative individuals able to deal with uncertainty.

The second staffing characteristic that was significantly associated with performance was *managers' tenure with an organization* ($p = .05$). Statistical analysis indicates that company tenure was associated with performance, suggesting that top performers' senior management teams are comprised of individuals with more company-experience than lower-performing units. Tenure with a company implies organizational experience which appears to have a positive effect on a manager's ability to "get things done." It is not surprising that managers' experience with an organization's procedures, goals, and personnel would be associated with greater effectiveness. Moreover, managers with more company experience presumably have better networks than their lesser-experienced counterparts, which could enhance information-processing capabilities.

In this study, we also asked the responding general managers to indicate the importance of traditional managerial skills for their top management team. We found it somewhat surprising that this management-skills measure was not significantly associated with organizational performance. A less rigorous statistical confidence level did reveal a relationship that was significant at the $p < .10\%$ level however.

Compensation

A second set of regressions focused on compensation practices: are there methods of paying managers that are more strongly associated with performance than other methods? A great deal of HRM research has addressed the issue of whether compensation affects individual performance (Milkovich & Newman, 1987); however, the linkage with organizational performance is unexplored. While the results of previous research on compensation/individual performance are inconclusive, assuming that employee performance *is* linked to how they are compensated, it is only a small step to extrapolate further and propose a link to organizational performance as well. In both business research and in practice there is a widely-held belief that a company's performance is a function of its managers' performance. If pay and incentive systems are successful in motivating managers to improve their own performance, it seems reasonable to assume organizational performance will be boosted as well.

Our questionnaire gathered data on two dimensions of executive compensation:

1. *bonus policy*—whether a bonus was paid, bonus amounts, the frequency of bonus awards, the extent to which bonuses are determined by formulas, and the role of individual performance in calculating bonuses; and
2. *salary particulars and philosophy*—base salary, a comparison with industry practices and salary equity within the top management team.

Of these eight variables, only the competitiveness of executive compensation as compared with industry practices was significantly associated with performance in the total sample.

The performance/salary competitiveness relationship ($p < .01$) indicates that executive compensation packages (including both salaries and bonuses) are more market-competitive in high-performing organizations. Paying managers more than the "going rate" may reflect companies' desire for quality managers, since paying above-market salaries has obvious implications for competence (Carroll, 1988). Thus, higher salaries may facilitate recruitment of exceptional managers. Furthermore, to the extent that compensation affects managers' motivation, attractive salaries and bonuses may affect management commitment to company goals, resulting in improved organizational performance. Finally, compensation may be used to encourage creativity and prudent risk-taking behavior in a challenging environment—characteristics that we have previously identified as beneficial to organizations.

Training

Data was collected on executives' participation in diverse training activities: formal (seminars and conferences) and informal (mentoring, required reading), in-house and external, specialized/functional and general management. Due to the high correlations among individual training measures, data on all 10 training activities were combined into an aggregate variable providing an overall measure of executive training participation. There was no statistical relationship between training and performance (however there was a relationship at the .10 level). (Individual regressions on more disaggregate training variables such as formal training and informal training also did not reveal any significant relationships.) These findings imply that in general top executive training may not have a discernable effect on a company's performance.

How might training's apparent lack of a significant relationship be explained? In evaluating the "worth" of training or any EHRM policy from an organizational perspective, one must consider the costs as well as the benefits. Costs must include not only the monetary cost of the training program, but the contribution of time required for managers' attendance. An hour spent in training is an hour foregone at some other activity, and at the senior-management level this "opportunity cost" may outweigh the benefits of training participation. Furthermore, if high-performing companies are recruiting managers with highly-developed skills to fill their executive positions (a pattern we detected in our sample), training may be less necessary. Alternatively, there could be personal development activities outside the work environment that serve to keep executive skills at high levels, compared to the more traditional training activities that formed the basis of our training measure. Finally, it is quite possible that executive training can impact only on long-term rather than the relatively short-term performance measure used in this study.

Performance Appraisal

The final EHRM policy area considered in this research is how managerial performance is evaluated and how this information is utilized. Six evaluation measures were studied including the frequency of appraisals (formal and informal), the use of objective and subjective criteria, the extent to which outcomes were emphasized, and the extent to which appraisal results are incorporated in broader EHRM policymaking. Of the six variables, four statistically significant relationships were identified. These included: informal appraisals of managerial performance ($p < .05$), relatively high-utilization of performance appraisal results ($p < .01$), and the use of both objective ($p < .05$) and subjective ($p < .01$) criteria of performance. These were all positively associated with organizational performance.

Greater frequency of performance appraisals implies an emphasis on the level of managerial performance, which may serve as an incentive to encourage sustained work quality. It is interesting to note that top performers use informal—rather than formal—performance appraisals more frequently than lower-performing organizations. This may imply less of an reliance on systematic methods that may be conducted on a "rote" basis, relying instead on a continual appraisal process driven by frequent contact between functional managers and the general manager as a basis for evaluation.

The positive relationship between the use of evaluation results with organizational performance indicates that top performing organizations take managerial evaluation seriously and use the results:

- to help determine merit increases and bonuses;
- to provide feedback on job performance;
- for job development purposes; and
- to help make staffing decisions.

These findings imply that performance appraisal systems are more fully-integrated with other EHRM subsystems (such as compensation) in high-performing organizations. Thus, the design of EHRM among these companies may be consistent with a systems approach, where each subsystem is interdependent, supportive and consistent with other EHRM subsystems. Theorists have promoted this mutually-supportive approach as maximally effective for the majority of organizations (Carroll & Schneier, 1982). The finding that the use of objective and subjective criteria of performance were related to performance may indicate that using multiple criteria of performance, even if done informally, is especially appropriate for top level managers.

These findings together suggest that performance appraisal systems appear to be more sophisticated in high performing companies; greater use of evaluation results involve tailoring criteria and methods to meet various end-uses. Furthermore, the use of appraisal results as inputs to other decisions suggests superior information gathering and processing capabilities. These results support the emphasis that some theorists have placed on performance appraisal (Carroll & Schneier, 1982; Devanna, 1984; Schuster, 1984) and suggest that careful evaluation of managerial performance may yield significant performance benefits for most firms.

EHRM's Universal Effect on Performance: A Concluding Note

We set out in this section to examine the linkage between organizational performance and EHRM and, in the process, identified six practices that are associated with higher levels of organizational performance:

- Emphasis on innovative/entrepreneurial personality characteristics when recruiting executives.
- Executive tenure.
- Competitive executive salary and bonus packages.
- Frequent evaluation of executive performance on an informal basis.
- Use of diverse (both objective and subjective)performance criteria when evaluating executive performance.
- Incorporating performance appraisal results into diverse EHRM policy decisions.

We speculate that these policies influence the skill level of top managers or their motivation, which can lead to improved individual performance, which in turn can have a positive effect on the organization's performance. Thus, we conclude that certain EHRM practices may "pay off" in the general business population by outperforming alternative policies.

The apparent linkage between individual EHRM practices and organizational performance lends support to a central aspect of the new "strategic orientation" to HRM; namely, that HRM (or in our case, EHRM) can affect a company's "bottom line." While regression analysis does not establish the causality which is needed to conclude that EHRM policies *lead* to higher performance, our findings represent a first step in identifying—and designing—more effective EHRM systems. In the following section we carry our analysis a step further to examine whether the effect of EHRM on performance is determined, in part, by the extent to which organizations are pursuing innovative strategies.

TESTING A CONTINGENCY THEORY OF EHRM: THE LINKAGE BETWEEN ORGANIZATIONAL PERFORMANCE, EHRM AND STRATEGY

Results

A moderating variable representing the interaction between the independent variables (innovation and EHRM) was added to each of the 19 regressions discussed. The results, which are summarized in Table 3, identified three (16%) statistically significant moderating relationships at the .05 level or above, two of which are in the compensation area. Discussion of the implications of these findings follows.

The main goal of this research project was to attempt to provide an empirically-based analysis of the contingency theories so popular in the business field today, which claim that the effectiveness of various HRM practices is related to their congruence with a particular strategic thrust of the organization. Focusing on HRM policies at the levels of executives only, we hypothesized that EHRM systems tailored to acquire and maintain the management resources and behaviors dictated by the organization's strategy would be related to firm peformance.

As described in the methodology discussion in mathematical terms, the proposed contingency theory states that strategy (in our case, innovation strategy) is expected to moderate the relationship between EHRM and organizational performance (i.e., innovation is expected to systematically modify the form of the relationship between dependent—organizational performance—and independent—EHRM practices—variables). A significant moderating effect implies that innovation interacts with EHRM to modify (improve or detract from) the EHRM/performance relationship found in the total sample. Moderated regression analysis goes beyond the issue of the EHRM/performance relationship addressed in the previous section to examine the *relative efficacy* of EHRM in different strategy environments.

Findings from our regression analysis indicates that of the four EHRM general practices we considered, only compensation and training policies or practices were linked specifically to performance in innovative organizations and these were in ways not expected. Thus, while our findings lend some mild support to a contingency theory of compensation or training, we generally found no basis for concluding that innovation strategy moderated the EHRM/performance relationship for most of the EHRM practices.

Table 3. Results of Moderated Regression Analysis New Product Innovation as a Moderator of the EHRM/Performance Relationship (Test of Proposition 2)

	Unstandardized Regression Coefficients		
Interaction Variable	*Coefficient a (Standard Error)*	ΔR^2	*F-Value*
Staffing			
Source of New Hires x New Product Innovation Strategy	−.127 (.111)	.016	1.311 (1.82)
Imp. of Managerial Skills x New Product Innovation Strategy	.046 (.105)	.002	.189 (1.87)
Imp. of Innovative Characteristics x New Product Innovation Strategy	.052 (.010)	.002	.222 (1.90)
Managers' Tenure x New Product Innovation Strategy	−.026 (.125)	.001	.045 (1.84)
Compensation			
Are Managers Paid a Bonus? x New Product Innovation Strategy	.006 (.077)	.000	.007 (1.93)
Bonus Based on Individual Performance x New Product Innovation Strategy	−.042 (.121)	.002	.122 (1.80)
Potential Bonus Amount x New Product Innovation Strategy	−.078 (.085)	.010	.857 (1.81)
Use of Formulas to Calculate Bonuses x New Product Innovation Strategy	.206*	.042	3.767
Frequency of Bonus Awards x New Product Innovation Strategy	.052 (.100)	.004	.268 (1.79)
Internal Equity of Salaries x New Product Innovation Strategy	−.081 (117)	.005	.006 (1.87)
Competitiveness of Compensation Package x New Product Innovation Strategy	−.265** (117)	.071	6.991 (1.87)
Performance Appraisal			
Frequency of Formal Appraisals x New Product Innovation Strategy	.096 (.092)	.012	1.092 (1.88)
Frequency of Informal Appraisals x New Product Innovation Strategy	−.037 (.101)	.002	.134 (1.86)

(*continued*)

Table 3 (Continued)

Focus on Outcomes x New Product Innovation Straegy	.008 (.135)	.000	.003 (1.84)
Use of Objective Performance Criteria x New Product Innovation Straegy	.067 (.101)	.005	.434 (1.87)
Use of Subjective Performance Criteria x New Product Innovation Strategy	.107 (.102)	.005	1.109 (1.86)
Utilization of Evaluation Results x New Product Innovation Strategy	.063 (.117)	.003	.288 (1.100)
Training			
Participation in Training and Development New Product Innovation Strategy	−.203* (.099)	.046	4.219 (1.86)

Notes: [a]The regression equations were run as:

$$P = a + bN + cX = dNx$$

where: P = Organizational Performance; N = New Product Innovation; X = EHRM Variable under consideration; and NX = Interaction between N and X. The variables P, N, and X were standardized prior to the regressions. Delta R^2 indicates the increase in R^2 due to the introduction of the interaction term NX into the equation.

*$p < .05$.

** $p < .01$.

Innovation as a Moderator of the Compensation/Performance Relationship

Moderated regression analysis indicated that of the seven compensation variables studied, there were two significant interaction effects. These findings can be interpreted to mean that certain compensation practices appeared to be more effective in certain new product innovative strategy environments than in others. The two compensation practices that seemed to "fit" better with particular new product innovation strategies were the use of formula or objective methods to calculate executive bonuses and the competitiveness of the top exceutives' compensation packages.

Thus, performance in firms making more frequent product changes was positively associated with use of *objective methods to calculate executive bonuses* (e.g., formulas, $p < .01$). The relationship between formula-based bonuses and organizational performance may be due, in part, to our innovation measure—new product development—which implies a strong marketing or sales focus. Typically, salary policies geared towards sales professionals have a large incentive component, which is often based on objective market data, namely sales. The success of a strategy promoting new products can be measured relatively easily with objective measures, particularly when compared to innovation strategies that focus on product quality or customer satisfaction. The availability of objective market indicators on which to base bonuses—and the relative importance of this type of outcome to companies pursuing strategies based on new product development—may make formulas a logical method for calculating bonus awards in innovative companies. The other significant compensation interaction relationship between competitiveness of the compensation package and performance, since it is negative, could be interpreted as indicating that market-based salary competitiveness is especially important in firms not involved in much product change: more stable and bureaucratic firms. Other motivational elements may be much more important than the competitiveness of the compensation package in high product innovation firms. Here the competitiveness of the compensation package may not be related to performance at all.

New Product Innovation as a Moderator of the Training and Development Relationship

As Table 3 indicates, the interaction term for new product innovation and amount of time spent in training and development for these executives is negative. This can be interpreted to mean that it is in companies which are low on new product innovation where the amount of executive participation in training and development activities has a positive relationship to performance. We do not know for sure what the relationship between participation in training and development activities and firm performance is for firms high in new product innovation—perhaps there is no relationship. It is hard to imagine that there is a negative relationship between performance and training in innovative firms, though this is possible since time spent in training and development might mean less time spent on more critical performance enhancing activities as we previously mentioned.

New Product Innovation as a Moderator of Staffing and Performance Appraisal Practices

No significant relationships were identified through our moderated regression analysis where specific staffing or performance appraisal practices were associated significantly with performance differently in companies which changed products frequently as compared to firms that change products more slowly.

Innovation Strategy as a Moderator of the Relationship Between EHRM and Performance: A Concluding Note

Increasingly, managers are being encouraged by academics to unleash HRM as a powerful management tool to help bolster business performance (Devanna, et al., 1984; Dyer, 1984a; Nkomo, 1980). This recommendation is based on the belief that strategies, in order to be successful, must be matched with appropriate processes, including HRM. At the heart of this argument is Galbraith and Nathanson's contingency theory of strategy implementation which theorists have adopted to link HRM directly to strategy and organizational performance (Devanna, Fombrun, Tichy, & Warren, 1982). Consistent with contingency theory is the belief that strategy dictates unique resource requirements (in this case, human resources or, more specifically, executive resources) which can be met by judicious use of HRM activities (Devanna et al., 1984; Miles & Snow, 1984). By "matching" HRM with strategy, critical human resources can be acquired, developed, motivated and retained. As a result, strategy implementation and performance will be enhanced (Foulkes, 1986).

Previous research has supported the strategy/HRM and HRM/performance relationships separately. However, to date, published research linking all three concepts has been limited to case analysis or small field studies. This study aimed to fill this void by attempting to provide these contingency theories of HRM some empirical support for linkages that previously were only theoretical or prescriptive in nature. Moreover, it was hoped that the relatively large sample size representative of all significant manufacturing industries would enhance the generalizability of the research findings.

However, since few contingency findings emerged, the study does not provide much support for a contingency theory of HRM. There appears to be more support for a normative theory of HRM. We must remember that we looked only at top executives and not at other organizational levels and there may have been a difference if other groups were studied; however we did expect the most likely contingency relationships would be found for top managers. Also the particular strategy measure here was new product changes rather than some other perhaps more critical types of strategy dimensions, at least in terms of their impact on HRM programs and activities. Failure to support a theory may be due to other reasons as well as the possibility of the theory being itself incorrect. Our measures of EHRM practices, performance, and new product strategy may not be adequate for testing a contingency theory of HRM even though we invested an enormous amount of time and thought into these measures. Field research is always subject to various restraints such as those factors impacting on response rates that force one to use less than optimum measures.

Another shortcoming of this research is that no causality is established. Conclusions that EHRM, innovation strategies, and performance are associated fall short of what is needed to develop an empirically-based, normative model. Case analyses comparing EHRM systems within high-performing businesses with those in low-performing businesses following similar innovation strategies would be helpful in further evaluating the EHRM/strategy/performance relationship, and may help establish causal links. Case analysis would be particularly useful in further analyzing the findings that seem counterintuitive: the apparent lack of benefit associated with management training, for example, or the apparent efficacy of compensation package competitiveness only in less innovative firms. Some follow up case studies are planned. Other ways this research could be expanded to provide deeper insight into "strategic" HRM include examining the HRM/performance link at other levels of the organizational hierarchy, inclusion of more specific measures of employee skills and talents, and broadening the HRM concept to include more practices and linkage with strategy formulation (in addition to implementation). Exploratory research on the interdependencies of HRM subsystems (e.g., the mutually supportive characteristics of performance appraisal and compensation systems) would make it easier to recognize and identify configurations of generic HRM "archetypes" and allow us to take into account the synergy within the HRM system, further improving our understanding of the real costs and benefits of HRM within business organizations.

APPENDIX

Operationalization of Study Measures

New Product Innovation Strategy: Respondents were asked what percentage of their business unit's total sales were accounted for by products introduced during the past 3 years.

Organizational Performance: Respondents were asked to indicate (1) the degree of importance that superiors attach to, and (2) their perception of the business unit's performance (relative to expectations) for each of the following dimensions of organizational performance: Sales growth rate, market share, operating profits, profit to sales ratio, cash flow from operations, return on investment, new product development, market development, R&D activities, cost reduction programs, personnel development, political/public affairs. The scale for importance ranged from "of little importance" to "extremely important," while the ratings ranged from "not at all satisfactory" to "outstanding." A weighted average was calculated using the level of importance as criterion weights.

R&D Importance: Rated from "of little importance" to "extremely important." Used as a construct validity check of new product measure.

Competitive Strategy: Respondents were asked to indicate what percentage of their business unit's sales are currently accounted for by products representing Porter's three competitive strategy types: low cost, differentiation, simultaneous low cost/differentiation. Used as a construct validity check of new product measure.

EHRM Variables: For each of the following measures, data was collected for 3 top functional managers which were averaged into a single "executive" measure.

Source of Management Hires: Respondents indicated for each top manager (R&D, Marketing, Manufacturing) the number of years in the current job and the number of years with the company. If years in the current job were equal to years with the company, the manager was hired from outside the firm.

Importance of Managerial Skills: For each top manager, respondents indicated the importance of various types of managerial skills in doing their jobs: technical expertise in functional field, general management skills, interpersonal skills, strategic planning skills, communication skills, time management skills, goal setting skills, skills in evaluating quantitative data, and persuasive skills. Each skill was rated on a 7-point scale ranging from "not important" to "extremely important." Data was aggregated into a single "management skills" measure.

Importance of Innovative Personal Characteristics: Same method used as for managerial skills, including the following characteristics: creativity, charisma/likeability, adaptability/flexibility, mental toughness, stress tolerance, initiative, willingness to take risks, and tolerance for ambiguity.

Company Tenure: Respondents indicated years spent with company for each senior manager.

Are Managers Paid a Bonus? Yes or no.

Potential Bonus Amount: Provided the numeric value of percent of base salary that could have been earned as a bonus last year.

Frequency of Bonus Awards: Indicated quarterly, semi-annually, annually or less often.

Use of Formulas to Calculate Managers' Bonuses: Are bonuses determined solely on the basis of formulas? Yes or No.

Bonus Amount Based on Individual Performance: Respondents indicated percentage of bonus based on the individual's, the division's, and the entire corporation's performance for a total of 100%. The individual component was used for this measure.

Internal Equity of Salaries: Because salary information is so sensitive, we asked respondents to indicate each manager's salary expressed as a percentage of the General Manager's salary. These percentages were compared—businesses where the three managers received similar salary levels rated received a low score on this measure which was translated to mean high equity. Large differences were on the inequity end of the spectrum.

Competitiveness of Compensation Package: Indicated on a 7-point scale how managers' compensation packages compared with compensation for managers in similar positions within the industry. Scale ranged from "above industry average by more than 20%" to "below industry average by more than 20%." Midpoint was the industry average.

Frequency of Formal Appraisals: Monthly, quarterly, semi-annually, annually, less often.

Frequency of Informal Reviews by General Managers: Same as above.

Evaluation Based on Actual Outcomes: Expressed as a percentage—on outcomes/results and process for a total of 100%.

Use of Objective Criteria: On a 7-point scale ranging from "not important" to "extremely important," indicated the importance that various types of performance indicators have had for each of the senior managers including: achievement of budgeted goals, meeting important deadlines, meeting quality standards, meeting quantity standards, new product development. Data was aggregated into single measure.

Use of Subjective Criteria: Same as above for development of subordinates, personal growth, process improvements, cooperation with others.

Utilization of Appraisal Results: Respondents indicated on a 5-point scale ranging from "never" to "always" how often the assessment of senior managers' performance was used for each of the following purposes: to determine merit increases, to determine performance bonuses, to provide feedback on job performance, for job development purposes such as training, to help make staffing decisions regarding promotions, transfers or termination. Data was aggregated into single measure.

Training Participation: Indicated senior managers' participation during the past 12 months in the following types of training activities: Formal (in-house and external; general management and functional area of expertise), systematic mentoring, special assignments, required attendance at conferences, required technical reading, required general management reading.

REFERENCES

Baird, L., Meshoulam, I., & Give, G. (1983). Meshing human resource planning with strategic business planning: A Model Approach. *Personnel, 60,* 14-25.

Carroll, S.J. (1987). Business strategies and compensation systems. In D.B. Balkin & L.R. Gomez-Mejia (Eds.), *New perspectives in compensation* (pp. 343-355). Englewood Cliffs, NJ: Prentice-Hall.

Carroll, S.J. (1988). Handling the problem of the need for consistency and the need for contingency approaches in the management of compensation. *Human Resource Planning, 11,* 191-196.

Carroll, S.J., & Schneier, C.E. (1982). *Performance appraisal and review systems.* Glenview, IL: Scott, Foresman.

Cook, D.S., & Ferris, G.R. (1986). Strategic human resource management and firm effectiveness in industries experiencing decline. *Human Resource Management, 25,* 441-458.

Dess, G.G., & Davis, P.S. (1984). Porter's generic strategies as determinants of strategic group membership and organizational performance. *Academy of Management Journal, 27,* 467-88.

Devanna, M.A. (1984). The executive appraisal. In C. J. Fombrun, N.M. Tichy, & M.A. Devanna (Eds.), *Strategic human resource management* (pp. 101-109). New York: Wiley.

Devanna, M.A., Fombrun, C.J., & Tichy, N.M. (1984). A framework for strategic human resource management. In C.J. Fombrun, N.M. Tichy, & M.A. Devanna (Eds.), *Strategic human resource management* (pp. 33-51). New York: Wiley.

Devanna, M.A., Fombrun, C.J. , Tichy, N.M., & Warren, L. (1982). Strategic planning and human resource management. *Human Resource Management, 21,* 11-17.

Dimick, D., & Murray, V. (1978). Correlates of substantive policy decisions in organizations: The case of human resource management. *Academy of Management Journal, 21,* 611-623.

Drucker, P. (1985). *Innovation and entrepreneurship.* New York: Harper & Row.

Dyer, L. (1984a). Studying human resource strategy: An approach and an agenda. *Industrial Relations, 23,* 156-169.

Dyer, L. (1984b). Linking human resources and business strategies. *Human Resource Planning, 7,* 79-84.

Evans, P., Doze, Y., & Laurent, A. (Eds.). (1989). *Human resource management in international firms.* London: Macmillan.

Fombrun, C.J., & Tichy, N.M. (1983). Strategic planning in human resource management: At rainbow's end. In R. Lamb (Ed.), *Recent advances in strategic planning* (pp. 319-332). Greenwich, CT: JAI Press.

Ford, J.D., & Schellenberg, D.A. (1982). Conceptual issues of linkage in the assessment of organizational performance. *Academy of Management Review, 7,* 49-58.

Foster, R.N. (1986). *Innovation.* New York: Summit Books.

Foulkes, F. (1986). *Strategic human resource management: A guide for effective practice.* Englewood Cliffs, NJ: Prentice-Hall.

Frohman, M.A. (1984). Human resource management and the bottom line: Evidence of the connection. *Human Resource Management, 23,* 315-334.

Galbraith, J.R., & Nathanson, D.A. (1978). *Strategy implementation: The role of structure and process.* St. Paul, MN: West.

Golden, K.A., & Ramanujam, V. (1985). Between a dream and a nightmare: On the integration of the human resource management and strategic planning processes. *Human Resource Management, 24,* 429-452.

Gupta, A.K. (1984). Contingency linkages between strategy and general manager characteristics: A conceptual examination. *Academy of Management Review, 9,* 399-412.

Gupta, A.K., & Govindarajan, V. (1984). Business unit strategy, managerial characteristics and business unit effectiveness at strategy implementation. *Academy of Management Journal, 27,* 25-41.

Gupta, A.K., & Taylor, M.S. (1989). *Impact of business unit strategies on executive career outcomes.* Symposium paper presented at the 49th annual meeting of the Academy of Management, Washington, D.C.

Heneman, H. G. (1974). Comparisons of self- and superior ratings of managerial performance. *Journal of Applied Psychology, 59,* 638-642.

Imai, M. (1986). *Kaizen.* Chicago: Random House.

Kanter, R.M. (1983). *The change masters.* New York: Simon & Schuster.

Kelleher, E.J., & Cotter, K.L. (1982). An integrative model for human resource planning and strategic planning. *Human Resource Planning, 5,* 15-27.

Kerr, J.L. (1982). Assigning managers on the basis of the life cycle. *Journal of Business Strategy, 2,* 58-65.

Labelle, C. M. (1983). *Human Resource Strategic Decisions in Response to Environmental Challenges.* Unpublished Masters Thesis, Cornell University.

Lawrence P. R., & Lorsch, J. W. (1967). *Organization and environment.* Boston: Division of Research, Graduate School of Business, Harvard University.

Lenz, R. T. (1981). Determinants of organizational performance: An interdisciplinary review. *Strategic Management Journal, 2,* 131-154.

Lorange, P., & Murphy, D.C. (1984). The human Resource dimension in the strategic planning process: System design considerations. In C.J. Fombrun, N.M. Tichy, & M.A. Devanna (Eds.), *Strategic Human Resource Management* (pp. 275-286). New York: Wiley.

Martin, M.J. (1984). *Managing technological innovation and entrepreneurship.* Reston, VA: Reston Publishing.

Miles, R.E., & Snow, C.C. (1978). *Organizational strategy, structure and process.* New York: McGraw-Hill.

Miles, R.E., & Snow, C.C. (1984). Designing strategic human resource systems. *Organizational Dynamics,* 36-52.

Milkovich, G.T. (1988). A strategic perspective to compensation management. In L. Rowland & G. Ferris (Eds.), *Research in Human Resources Management* (pp. 263-288). Greenwich, CT: JAI Press.

Milkovich, G.T., & Boudreau, J.W. (1988). *Personnel/human resource management.* Plano: TX: Business Publications.

Milkovich, G.T., & Newman, J. (1987). *Compensation.* Plano, TX: Business Publications.

Misa, K.F., & Stein, T. (1983). Strategic HRM and the bottom line. *Personnel Administrator, 28,* 27-30

Nkomo, S.M. (1980). Stage three in personnel administration: strategic human resources management. *Personnel, 57,* 69-77.

Olian, J.D., & Rynes, S.L. (1984). Organizational staffing: Integrating practice with strategy. *Industrial Relations, 23,* 170-183.

Perry, L.T. (1984). Key human resource strategies in an organizational downturn. *Human Resource Management, 23,* 61-75.

Peters, T.J., & Waterman, R.H. (1982). *In search of excellence.* Cambridge, MA: Harper & Row.

Porter, M.E. (1980). *Competitive strategy.* New York: Free Press.

Schuler, R.S., & Jackson, S.E. (1987a). Organizational strategy and organizational level as determinants of human resource management practices. *Human Resource Planning, 10,* 143-56.

Schuler, R.S., & Jackson, S.E. (1987b). Facilitating entrepreneurship to gain competitive advantage: Implications for human resource management practices. In F. Hoy (Ed.), *Proceedings of the 47th annual meetings of the Academy of Management* (pp. 274-280). New Orleans, LA.

Schuler, R.S., & MacMillan, I. E. (1984). Gaining competitive advantage through human resource management practices. *Human Resource Management, 23,* 241-255.

Schuster, J. (1984). *Management compensation in high technology companies,* Lexington, MA: Lexington Books.

Smith, E.A. (1982). Strategic business planning and human resources: Part 1. *Personnel Journal, 61,* 606-682.

Souder, W.E. (1987). *Managing new product innovations.* Lexington, MA: Lexington Books.

Stonich, P.J. (1981). Using rewards in implementing strategy. *Strategic Management Journal, 2,* 345-352.

Szilagyi, A.D., & Schweiger, D.M. (1984). Matching managers to strategies: A review and suggested framework. *Academy of Management Review, 9,* 626-637.

Tichy, N.M., Fombrun, C.J., & Devanna, M.A, (1984). The organizational context of strategic human resource management. In C. J. Fombrun, N.M. Tichy, & M.A. Devanna (Eds.), *Strategic human resource management.* New York: Wiley.

Tomasko, R.M. (1982). Focusing company reward systems to help achieve business objectives. *Management Review, 71,* 8-12.

HUMAN RESOURCES STRATEGIES FOR LATERAL INTEGRATION IN HIGH-TECHNOLOGY SETTINGS

Susan A. Mohrman, Allan M. Mohrman, Jr., and Susan G. Cohen

INTRODUCTION

The literature on high-technology management points overwhelmingly to the importance of lateral integration processes for successful performance. This chapter reviews the factors that make lateral integration so important, briefly discusses various approaches to achieving such integration, and then concentrates on the key organizational design factors and the characteristics of the human resources management systems that foster lateral integration.

THE IMPORTANCE OF LATERAL INTEGRATION

High-technology organizations exist in highly complex and turbulent environments. Such environments have been found (Lawrence & Lorsch, 1967)

Human Resource Strategy in High Technology.
Advances in Global High-Technology Management, Volume 1, pages 31-57.

ISBN: 1-55938-489-1

to pose requirements both for enhanced differentiation and for greatly increased integrative mechanisms to ensure collaboration of subunits whose tasks must be integrated around the completion of the global task.

High-technology firms must process considerable information to cope with uncertainty stemming from environmental turbulence and inherent in the nature of high technology tasks. Galbraith (1973) developed a framework for understanding organizational design based on the organization's information processing needs. In an organization where the work is complex and uncertain, the standard coordinative mechanisms—goal-setting, hierarchy, and rules and programs—must be supplemented by other approaches. Two of these, the creation of self-contained teams and the use of lateral coordinating mechanisms, are lateral in nature. The third approach, the development of vertical information systems, bolsters the ability of the hierarchy to integrate by providing decision makers higher in the organization with more timely information. The fourth, the creation of organizational slack (performance slippage) happens by default when coordination is not fully achieved. These approaches to integration are not mutually exclusive; rather, a complex organization that cannot allow performance to slip has to rely heavily on lateral coordination mechanisms to supplement vertical integrative approaches.

The *extreme performance pressures* that confront high-technology firms require lateral approaches to integration. The competitive environment in fields such as electronics, pharmaceuticals, and biotechnology places demands on firms to keep up in rapidly developing technological arenas, achieve breakthroughs, and rapidly introduce new products that are of high quality and can be manufactured efficiently. Slack is no longer an option: companies are required to compete on schedule, cost, quality, and technology development in order to survive.

A great deal of work has been done that points to the need for greatly increased lateral integration in order to address these performance pressures. Pressures for quality have led to widespread implementation of quality improvement frameworks (Crosby, 1979; Deming, 1986a, 1986b; Juran, 1964, 1988; Juran & Gryna, 1980). These approaches emphasize the need for breaking down both vertical and horizontal organizational boundaries that serve as barriers to the lateral, cross-functional collaboration necessary to improve organizational processes and enhance quality. They stress that quality is best defined as responsiveness to the requirements of both external and internal customers rather than to standards that emanate from the organizational hierarchy. This approach underscores the importance of close working relationships between design, manufacturing, and marketing as the organization struggles to remain in touch with and responsive to the external market, and as units orient their activities to meeting the needs of the internal customers who receive their work and with whom they are interdependent.

The quality improvement frameworks' emphasis on organizational processes that weave across different units of the organization and their orientation to customer force a lateral view of the organization and its integrative needs. These frameworks identify "continuous improvement" of organizational processes as the appropriate organizational goal. The systematic analysis of processes and introduction of changes to improve them is accomplished by the people actually carrying out the processes. This continuous improvement process itself introduces substantial information processing requirements into the technical core of the organization and increases the needs for lateral integration.

The performance pressure for *speed* manifests itself in demands for significantly more rapid new product development and time to market cycles, and for on-time delivery (Stalk, 1988). Most treatments of new product development mention cross-functional teamwork as a key element in achieving fast cycle time (e.g., Bower & Hout, 1988; Souder, 1988; Takeuchi & Nonaka, 1986), not only because of the need for parallel processing such as simultaneous engineering and for real-time resolution of coordinative needs, but also because of the learning that is required by organizations in order to introduce new products quickly.

The pressures for speed also exist at the strategic level, where a quickly changing competitive environment requires highly effective teamwork at the top. Top executive teams must be capable of making rapid decisions using extensive real-time operational and environmental data. In the "high velocity" high-tech environment, effective top executive teams use decision processes that enable key stakeholders to contribute to strategic decisions, but maintain the ability of the CEO to make timely decisions (Eisenhardt, 1989). Executive team members need to quickly define the implications of strategic decisions for their areas and begin implementing them. Thus, the lateral integration process begins at, and is modelled by, the top executive team.

Cost pressures require that organizations implement a number of "breakthrough" manufacturing techniques (Haas, 1987), many of which require interfunctional collaboration. The effective utilization of CAD/CAM, for example, promotes information processing across the various units involved in the-design-to production process, but also requires collaboration among these units in the development and implementation of the CAD/CAM system itself and its day-to-day utilization (Adler, 1990).

In addition to the market driven performance pressures being experienced by high-technology firms, the *technology* of these firms requires extraordinary amounts of information processing in the technical core of the organization, and cries out for lateral integration (Mohrman, Mohrman, & Worley, 1990). Technical organizations tend to involve numerous technical specialties whose work needs to be coordinated in service of an overall product. Rapid technological advances in the tools that these organizations utilize in doing their work, in the work processes they utilize, and in the products they produce, introduce a great deal of technical uncertainty that requires on-line problem solving. These very tools include powerful cybernetic, information processing mechanisms that make possible levels of real time lateral integration that permit parallel processing and interfunctional problem solving that were heretofore unthinkable. Network technology simultaneously improves both centralized control and decentralized lateral decision making (Applegate, Cash, & Mills, 1988). The development of computer software to support group and cooperative work is burgeoning (Johansen, 1988). This very capability has created its own demand, a fact which will no doubt change the nature of organizations irreversibly and move us toward a highly lateral image of organizational functioning.

Both technological complexity and relentless competitive environments place constant demands on high technology organizations for *innovation, the solving of complex problems, and organizational learning*. Literature on these processes stress that they are the product of teams. They involve novel combinations of perspectives and the application of various knowledge bases to issues, problems, and needs that cannot be effectively addressed within one existing frame of reference (e.g., Kanter, 1983; Mohrman et al., 1990; Pinchot, 1985). Addressing today's performance challenges, for example, requires a simultaneous juggling and resolution of the logics that have dominated how organizations think about cost, schedule and quality, generally requiring collaboration between various functions that have historically been tasked with "ownership" of one of these concerns. As organizations find that simply becoming more efficient in their current processes is an insufficient response to the challenges that they face, they will increasingly need to learn new ways of operating—innovative breakthroughs that go beyond incremental improvements. Both the conceptualization of these breakthroughs and the implementation of change entail lateral integration.

ORGANIZING FOR LATERAL INTEGRATION

Lateral integration is required in a high technology organization, not only for effective performance of the core technical tasks, but also to successfully link the organization to its marketplace and to ensure that it can successfully adapt to a rapidly changing technological and commercial environment. Organizations that exist in such a fast paced environment cannot rely on relatively slow movement of decisions through hierarchical levels to determine appropriate operational responses to environmental and market changes, to resolve technical uncertainties, or to mediate disputes between units that must collaborate in order to produce a product. Even with the existence of sophisticated vertical information systems, the time it takes for an overloaded upper management to process the information and draw and communicate appropriate organizational implications is a bottleneck in organizational operations. The useful contribution of rules, standards, programs, goals, and targets, although not unimportant, is diluted by rapid change that requires their continual reassessment.

Since slack is not a competitive option, high technology organizations must rely on a combination of the use of self-contained teams and sophisticated lateral integration techniques to extend their information processing capabilities (Galbraith, 1973). The former bound information processing by including all necessary functions in teams with well delineated tasks. Examples include product development, project and program, and customer service teams. The latter approach creates a network environment in which widely dispersed individuals and organizational units are connected by techniques including face-to-face (or technologically facilitated) interaction, the creation of special integrating roles, task forces and temporary teams, and the creation of structures such as matrix organizations that build in lateral coordination. Network environments are flexible and adaptable, permitting connections to be made or broken, strengthened or weakened, in response to change in business conditions and strategy (Savage, 1990). Network members must be able to cooperate across organizational boundaries as they apply their specialized expertise to cooperative tasks. Teamwork is an extremely important success factor in organizations that rely so heavily on lateral processes.

Extensive interviews that we have conducted in studies of high-technology organizations (Mohrman et al., 1990) have shown that organizational members generally appreciate the importance of these lateral processes. Indeed, these organizations tend to utilize lateral integrating devices more frequently than their low technology counterparts. On the other hand, much frustration exists about how well lateral integration is actually achieved.

The lack of teamwork and cooperation has been cited as a competitive disadvantage in American firms (Birnbaum-More & Nadler, 1989; Dertouzas, Lester, & Solow, 1989). For example, quality improvement proponents such as Deming (1986a, 1986b), innovation scholars such as Pinchot (1985), and students of time to market such as Bower and Hout (1988) have all advocated that organizations overcome the obstacles to effective performance that are inherent in the rather strong boundaries between functions in American organizations. Even in high-technology organizations, where there is an intuitive sense of the need for integration, there appear to be powerful factors in the way we organize that work against teamwork and integration and promote suboptimization.

Two design approaches that form the underpinnings for the traditional American way of organizing work against lateral integration. These are the proclivity toward *analytical* images of the organization and the emphasis on *hierarchical control.* Although it has been recognized that the nature of the environment and the work in high-technology firms is not fully amenable to these approaches (Anderson & Kleingartner, 1987; Dornbush & Scott, 1977), as high technology firms have grown and become more complex, these organizing principles have come to predominate.

Organizations have typically been structured analytically, separating different expertises, specialties and ways of looking at the world into different units, which are further broken down into like groups to provide easily supervised units with a critical mass of people performing similar work, and into jobs, which specify the particular set of tasks an individual is responsible for. The process of taking complex technical problems and breaking down the work and assigning pieces of it to units and then to individuals, often called "work breakdown," is core to the organization of most technology firms. Primary emphasis is on breaking apart and making work manageable rather than on integrating diverse perspectives. The creation of specialized staff groups is one manifestaion of the analytical model.

The emphasis on hierarchical control is embodied in the distinctions between the technical core of the organization which performs the work, the top management group which addresses strategic directions and sets targets, objectives and policy, and the often sizable group in the middle whose responsibility is to control the work of the technical core, and translate the strategic direction developed at the top into goals and processes for the technical core (Thompson, 1967).

Both the analytical organizing framework and the emphasis on hierarchical control work against lateral integration. The more work is packaged for highly differentiated units, the less easy it is to achieve integration and to bring to bear multiple perspectives in real-time on-line problem solving and coordination. The more the control function is vested in a special hierarchical group, the less those performing the work will be able to make on-line adjustments and to take responsibility for the iterative coordination processes that are required to work out complex interdependencies.

The argument here is not that analytic breakdowns and hierarchical control are unnecessary. Rather, it is that such design approaches have not enough been tempered and offset by systemic design mechanisms, resulting in dysfunctional segmentation of the organization (Ackoff, 1981; Dertouzos, et al. 1989; Kanter, 1983). Analytic and hierarchical organizational and human resource management approaches have been well-developed both in literature and practice. Less attention has been paid to organizational approaches to the synthesis of various parts of the organizational system and to building lateral processes by which people can control themselves.

Achieving the lateral integration necessary to address stringent performance requirements, effectively integrate a myriad of technical expertises, and solve complex problems that require the application of multiple frameworks requires ways of organizing that emphasize *systemic approaches* and that foster ongoing *self-design* by units throughout the organization that have to change their activities continually to adapt to strategic redirections and to marketplace and technological change (Mohrman & Cummings, 1989). The next section addresses some of the organizational design factors that contribute to the integrative and self-designing capabilities of organizations. The last section examines the human resources practices that are congruent with and promote integration and self-design in organizations, and that support the necessary design features.

KEY DESIGN FEATURES REQUIRED TO MANAGE LATERALLY IN DYNAMIC HIGH-TECHNOLOGY ENVIRONMENTS

The design features described below are required to enable the integrative and self-designing capabilities of organizations. They foster a systemic approach and more reliance on self control and less on hierarchical intervention.

Proliferation of Teams and Teamwork

Team design approaches are a key structural embodiment of lateral integration of interdependent employees, and the need for them has been discussed in some detail earlier. What needs to be emphasized here is that teaming mechanisms take many forms. Some are permanent work units; others are temporary structures such as product development teams that reconfigure as projects end and others begin; still others, such as quality improvement teams, are temporary improvement structures whose function is to analyze and improve work processes. An increasingly prevalent phenomenon are hierarchies of teams, in which cross functional management teams provide a strategic umbrella and an efficient approval process for cross-functional work teams whose members may have diverse functional reporting relationships. Finally, there are task focused networks: constellations or matrices of people that may lace throughout the organization and beyond and which are linked together by having a contribution to make to a common task. Such networks are characterized by teamwork but they are not teams with known or well-defined boundaries.

Executive Team Leadership

Individuals and units within an organization have difficulty integrating their activities if there is not a clear strategic direction and well-articulated goals and priorities that provide criteria for multifunctional decision making (Eisenstat & Cohen, 1990). Because of the changing environment, organizations must be nimble—able to change strategic direction as needed, and to quickly commit the various parts of the organization to these new directions (Eisenhardt, 1989). This requires executive-level teamwork. Executive decision making defines the context for lower-level cross-functional decisions. Top management makes strategic choices about what products to produce, technologies to develop, how to structure the organization, how to allocate resources, and how to approach the competition (Ancona, 1990). Without the capacity to integrate different functional perspectives and to resolve conflicts at a strategic level, suboptimization will occur at lower levels. Collaboration and integration at the top provides an umbrella and model for such activities throughout the organization, and provides broad decision guidelines and criteria.

Unencumbered Decision Making: Flatter Organizations, Empowered Teams

Lateral mechanisms can effectively reduce uncertainty only if they are empowered to make decisions within the broad strategic direction, guidelines and priorities specified by the executive team. If the decisions that emerge from lateral interactions must travel through multiple levels of several functions for approval and/or to arrive at a tie breaker, then the need for vertical information processing has not been reduced. Thus, a flat organizational structure, with readily accessible tie breakers is a requisite to speedy response. Teams may have to report to teams in order for approval of team decisions by all necessary stakeholders to occur in a timely fashion. Wherever possible, all necessary resources and authority should reside within or close to the operational levels that are continually altering their course as they work out uncertainties, interdependencies, and respond to change.

Less Compartmentalization

Traditional organizations have segmented parts of the organization that house different knowledge bases and perform different steps in the work flow. People who are housed in different functions are often highly interdependent, despite their organization segmentation. In organizations excellent at new product development and rapid cycle times, the organization is both flatter *and* less compartmentalized (Dertouzos et al., 1989). This has been achieved in some organizations by the creation of "super functions" (Galbraith, 1990) that house previously separate disciplines such as design engineering and manufacturing process development in a "New Product Development" function.

The segmentation of staff expertises such as human resources, planning, and quality break apart aspects of functioning from the core operations of the organization. Quality becomes the responsibility of people who are not performing the operations whose quality is in question. Human resource management becomes the responsibility of people who are not managing people. This violates the systemic quality of an organization and leads to rather absurd situations where "managers" feel responsibility for operations but not for managing the people who are performing them, or for productivity but not for the quality of the goods produced. It also leads to debilitating conflict between functions, each of which is suboptimizing around its own limited set of criteria. Restoring the systemic nature of the organization's work not only requires better lateral relations between functions, but also requires the recombination of responsibilities that cannot be broken apart. The current trend to return quality responsibility to the operations people is one manifestation of such combination of staff and line tasks.

Organizing Around Processes

Organizational structures that group people doing similar work and "separate out" specialties from the operational core of the organization often cut across the natural processes of the organization for reasons of efficiency and administrative control. Quality improvement experts point out that this makes it difficult to integrate the various components of a complete process and introduces error into the process flow. For example, one of Demings's 14 points is to "break down barriers between departments" (Deming, 1986b). Process integration is essentially lateral integration, and requires that people and units involved in different aspects of a process collaborate in the improvement of that process. Collaboration not only enables waste and inefficiencies to be identified and eliminated, but helps to create synergy and increase motivation for continuous improvement.

Rich Information Environment

If operational units are to be able to cooperate laterally to effectively integrate work, make good decisions, and make adjustments in the way they function, they must be privy to relevant information. A rich information environment is required, including both global shared data bases and local data bases that provide data pertinent to particular processes, products and customers. Teamwork requires real-time, credible data (Gupta & Wilemon, 1988) that enable team members to transcend their own opinions in order to make criterion based decisions, and to learn together how to improve their operating results and adapt to changing circumstances. Faster and more effective decision making occurs when more, not less, real-time information is used (Eisenhardt, 1989).

Rich Linkages to Customers, Suppliers, and other Stakeholders

Key competition in the marketplace has led increasingly to the recognition that the health of the firm depends on meeting the requirements of customers, the ultimate recipients of the products or services produced. This same concept has been extended internally, where it has been recognized that failure to meet the requirements of internal customers—those who receive the products and services of other parts of the organization—causes costly delays, expense rework, and poor quality as well. In a complex network, such internal customers may not be easily conceptualized as the next person or unit in the process: rather, complex patterns of interdependence mean that there are stakeholders throughout the organization who are affected by decisions, products, and services of any person or unit. Rich linkages, opportunities for mutual influence, and smoothly working interfaces are required for integration.

Computer mediated network technology permits links to be formed between interdependent contributors who are geographically dispersed. Electronic mail and video conferencing capability enable networks to be rapidly formed as "virtual organizations" without having to invest in physical space or administrative support. These virtual organizations can be rapidly dissolved when work is completed (Savage, 1990). However, more effective collaboration will occur if the network builds upon the informal organization. People are more likely to collaborate with those they know, respect, and trust. Advice will be sought from those who are known to be influential, knowledgeable, and supportive. Personal relationships are the glue that solidify network links.

Ongoing Redesign

Organizational design has typically been hierarchically determined. Changes in design have been labelled "reorganizations" and are seen as discontinuities in the life of an organization. In a laterally integrated organization, interdependent people will be continually rearranging their tasks and the way they do business as they adapt to the task at hand and to changes in market and technology. In a rapidly changing environment, building in this flexibility is key to the adaptive capabilities of the organization (Mohrman & Cummings, 1989). Successful high-tech firms have been found to be able to rapidly and frequently change their structure and to be able to communicate new structures effectively so that organizational members can quickly readjust their activities (Schoonhoven & Jelinek, 1990).

Flexibility requires authority to be distributed widely throughout the organization. Self-managing contributors become responsible for self-design by forging new links with other self-managing contributors, discontinuing other links, and making sure that appropriate problem solving and information exchange occur. Those with expertise are not only empowered to make business decisions; they become responsible for designing the forums within which decisions are made.

Although organizations and the organizational literature recognize that increased lateral integration is needed and that the design factors listed above are desirable, our studies suggest that there are powerful forces working against managing organizations in this manner. One key issue is that careers of individuals are greatly influenced by human resource practices that do not support lateral integration and change. Often teamwork is perceived as antithetical to self-interest as shaped by the functional and hierarchical systems of the organization, and by the human resources systems that conform to their logic. The next section will enumerate the qualities of human resources practices that support a systemic, dynamic view of the organization.

WHAT HUMAN RESOURCE PRACTICES ARE REQUIRED TO MANAGE LATERALLY?

The wide reaching study of productivity and competitiveness of American industry by the MIT Commission on Industrial Productivity identified the need for the introduction of innovative human resources approaches as a key focus for productivity improvement. Also recommending various restructuring efforts and the introduction of technology innovation, they concluded that "without changes in human-resource practices, the benefits of the other restructuring efforts will remain elusive" (Dertouzos et al., 1989, p. 126). The key importance of human resource practices also characterizes the quest by high-technology organizations to establish structures, introduce technologies, and improve processes of lateral integration.

Much press has been given to nontraditional human resource practices in start-up, high-technology firms. It has been our observation, however, that as these firms grow in size, the need to "rationalize" human resource practices emerges, and often the formal human resource systems of high technology firms begin to resemble those found in larger, more staid corporations. The danger in this comes from the fact that these traditional human resources systems were designed to meet the needs of organizations dominated by analytical and hierarchical logic and structures.

This section attempts to answer the question: What kind of human resources systems optimally fit the emerging designs that promote the lateral integration necessary to meet high-technology performance demands? The human resources systems that we consider are typically designed and administered by the Human Resources function, but of course their execution depends on the behavior of the employee base at large.

One set of systems are the "performance management" practices (i.e., the various ways by which the performances of people, whether individuals or groups, are defined, planned, and reviewed, and the processes by which performance capabilities are developed and improved). Performance management systems include goal-setting, appraisal, and training and development. Another set of systems guide the movement of people through various roles in the organization. These include career progression, succession planning, placement, selection, and development expriences that prepare people for their next jobs. A third set, reward systems, address the mechanisms by which value of contribution is established and rewarded, and equity perceptions are developed. Finally, we discuss the human resources information system, including data bases and tools to facilitate the administration of the other human resource systems.

In traditional organizations, designed analytically and bureaucratically, and embodying principles of hierarchical control, formal human resource systems are based on these same principles. Clear job descriptions and job evaluations based on market worth and position in the hierarchy form the basis for rewards and performance management practices in which the individual is the main focus. Performance management cycles conform to the calendar year, and individual performance is assessed independently of the natural cycle of work or the performance of the interdependent network in which the performer is enmeshed. In fact, elaborate systems such as ranking procedures are often used to discriminate among performers who perform similar jobs. Responsibility for the execution of the performance management systems lies with the supervisor. Training and development activities are designed to provide particular skills and knowledge bases to perform specific jobs.

These practices are inadequate for the fast paced, highly interdependent world of high technology firms (Mohrman et al., 1990). They are not well-suited for organizations dealing with the performance pressures described earlier that require organization designs that foster lateral integration, a more systemic view of performance, and self-designing capabilities.

Dimensions of Integration

In high technology settings achieving integration is extraordinarily difficult because the organization is at the same time highly differentiated (Lawrence & Lorsch, 1967) and highly interdependent. Thus, although the organization can and must be broken down into parts, the interdependence assures that it is not safe to assume that if the task of each part is done to specification then it will integrate into the whole smoothly. Essentially this means that all human resource practices at the level of the part, instead of being designed as if in a vacuum, must be done in the context of the whole.

Achieving synthesis, or integration, involves several dimensions. First, there are various *systemic levels.* At each level, the performing elements are nested in higher systemic levels. The sine quo non of managing synthetically (as opposed to analytically) is a notion of what the whole performance is toward which the parts are being managed. There are several levels of parts and wholes within an organization, from individual employees, through teams and larger organizational units (departments, product lines, programs, divisions), to the organization as a whole. Each level is a social entity made up of human resources. At each level it is possible to conceive of managing the human resources at that level of aggregation. In high technology settings the importance of synthesis mandates that it be done explicitly.

To acknowledge the systemic qualities of an organization, human resources must be managed at all levels: individuals, teams, larger units, and organizations. Practices must be developed that take the management of human resources in units larger than the individual seriously and that place the management of each level in the context of higher systems levels in which it is embedded.

One complication in managing multiple systemic levels is that there often is not a strict hierarchy in the nesting of individuals, teams, and the organization. Teams of people may overlap with one another, people may belong to more than one team, and teams might even stretch beyond organizational boundaries but include only part of the organization. This complicates the human resource management task, but underscores the importance of the next dimension, the management of lateral interaction between elements in the system.

The management of *interaction between elements in the system* is in essence a lateral process. Human resource management in high technology must reflect the criticality of interactions of various performing elements (individuals, teams, departments) with a potentially very large lateral role set, including customers, suppliers, and other interdependent performers. Traditional assumptions that the knowledge, skills, and responsibility for managing people lie primarily in the hierarchical reporting relationships are outmoded. Performance information lies in a lateral role set and human resource management practices must actively involve these other elements of the system. A possible complicating factor with this dimension is that lateral role sets cut across systemic levels and across hierarchical levels. An individual may deal with other individuals at the same or different hierarchical levels in the organization, with other teams, or with entire customer or supplier organizations. Teams may deal with other teams, with individuals, or with larger units.

The other dimension of integration is *time*. The dynamic external market environment and rapidly changing internal and external technological environments mandate ongoing learning and change. Human resource management practices must be flexible enough to respond to changing conditions; they must establish an expectation in people that change will occur.

Traditional human resources practices are not flexible. They are based on an analytic breakdown of the organization into jobs and responsibilities that may need to rapidly change, and on goals and objectives that may become irrelevant to a changed environment. Furthermore, the cycles of execution of the human resource practices generally reflect administratively convenient timetables such as the calendar year. The time cycles for interdependent high technology tasks are frequently more rapid (or much longer, as in the case of complex programs). Human resources processes should be designed to occur as needed in order to be responsive to the rate of change and the natural work cycles of the work that is being performed.

Having identified these three important dimensions of integration, we will next examine their implications for various human resources management systems. In the following sections we treat the major leverage areas of performance management, career development and human resource planning, rewards and compensation, and human resource information systems (HRIS). We describe each in terms of how the practices fit the central design problems in high technology organizations: systemic versus analytic practices, lateral versus hierarchical practices, and dynamic versus static practices.

Performance Management

Performance management is a set of human resource management activities that are aimed at directly affecting performance. There are three main components of performance management:

- defining and planning future performance;
- developing performance capabilities; and
- reviewing performance (Mohrman et al., 1990).

These components to performance management can happen at any time, whether or not performance is simultaneously taking place, and in any sequence, but most often the components are thought of as taking place in the order just listed. Reward systems are an integral part of performance management, but because they also serve other functions, they will be discussed separately in this section.

The high-technology organization of the future will have to manage performance at several systemic levels:

- the organization itself;
- teams and focused networks within the organization; and
- the individual employee.

The key to successful performance management lies in how well the organization is able to integrate the processes that occur at these various levels.

Presently, most organizations focus their performance management efforts at the individual level and conduct them in a hierarchical manner. Individuals are assigned goals and objectives, individuals are trained and developed (organization development is the exception here), and the performances of individuals are reviewed and evaluated by their supervisors. There is an analytic assumption that managing the performance of individual employees will aggregate automatically to positive performance at the organization and team levels. There is not much evidence for this. Our research in high-technology firms indicates that these individual performance management practices do not even significantly lead to individual performance (Mohrman et al., 1990). They have at best no effect at the team level and sometimes even have negative effect when they work to pit interdependent team members against one another (Deming, 1986a; Mohrman et al., 1990).

Traditionally, performance management has been a hierarchical responsibility (i.e., supervisors have been responsible for managing the performance of their subordinates). Research evidence has shown that hierarchical performance management of the individual does not affect performance. In the high technology organizations we studied we found that when team members participated in self management of their team's performance, higher performance at both the team and the individual member level resulted. In other words, performance management at the team level set the stage for productive performance management at the individual level. Team self-management of performance affected at least three things that were related to better individual performance: the amount of teamwork, the development of necessary individual skills, and the understanding individuals had about how their roles fit into the bigger picture of the team's task and mission.

Other research that we are presently doing on high technology teams indicates that there is a similar need for the team's performance to be managed within a process of performance management at the organization level. When teams, such as new product development teams, focus solely on their own performance abstracted from organizational issues, they frequently find their efforts drowned out or counteracted by developments at the organizational level. Lack of performance management at the organizational level most often results in uncoordinated functional decisions that impair cross-functional team performance. Changes in organizational direction and priorities that are not articulated with team performance management can overnight turn projects on the road to success into wastes of resources, or can result in crucial resources for one project being diverted to another without proper planning at the team level and without proper input from the team level.

Performance management at any lower level will not have an effect at that level or any level above it unless it is done within the context of performance management at the next higher level. The converse of this proposition is that when performance management is done in a participative way at a higher level it can have a positive effect at both that level and the next lower level providing that its own performance context is clear. Successful high-technology organizations will have to create performance management processes that operate at these multiple levels and integrate them.

The next stage of performance management is to develop the performance capabilities. Again, development of team capabilities sets up demands for organizational development and will set the stage for necessary individual development. Creation and maintenance of cross-functional teams involves development of the larger organization such as: organizational role arrangements, behavioral norms, application of distributed information technologies like CAD systems, and the like. These have implications for the capabilities that team members must have or develop. Several firms we have worked with have mentioned that these organization development tasks cannot be done once and for all, but require revisiting as new teams with new needs are formed.

Just as important, cross-functional teams at the operational level work best when higher level cross-functional management teams create a laterally integrated context of support for the project teams. It is this organizational level structure that helps accomplish the organizational level performance planning mentioned above, and that makes contextual decisions such as the funding of capital expenditures required for development of distributed information technologies to support operational self-management. These higher level teams must also be developed as effective performance units.

Performance reviews also must occur at multiple nested levels of analysis. We have noticed that if a team looks at its own performance as a team and discusses it in a problem solving way, an unavoidable byproduct is increased awareness of the contributions of the team members. Therefore, team performance review contains individual reviews within it. Similarly, both team and individual performance reviews invariably uncover contributions to performance beyond control of the target performers that stem from the larger system to which the performer belongs. In a sense, performance reviews at lower levels always contain reviews of the performance of the whole system to which it belongs. An engineer's failure to meet a scheduled milestone might be due to the fact that the team had communication problems, for instance. These realizations that come from multilevel reviews are the kind that Deming (1986a) sees as necessary starting points for continuous performance approval.

The high technology firm of the future will not only manage performance as nested systems but it will actively involve lateral role sets in the process. Most high technology work is done laterally. Traditional supervisors often have only a vague idea what the performances of employees are (Von Glinow, 1988). Individuals will link directly with those who are interdependent with their work. It is these people who have the necessary information for judging performance. For this reason lateral performance management practices will come to rely heavily on input from teammates, peers, customers, and suppliers. This argument applies to both team and individual levels of performance management. Teams will use customer needs and expectations as inputs to plan their performance and customer assessments to judge it.

The dynamic nature of the high technology organization puts demands on the performance management practices. Practices that rely on the assumption of stability will not work in these environments. We have found that fixed performance standards and goals have no effect on performance in high technology organizations (Mohrman et al., 1990). People resist having their performance managed by them and see them often as irrelevant. Goals are effective in facilitating performance when employees have ownership over them and when they are seen as redefinable as changing circumstances warrant. When strategic direction of the organization changes it will be important for teams to manage their adjustment to those changes. These things imply that performance management in high technology will be an ongoing process, as needed, rather than occurring on a predetermined administrative cycle.

The high-technology company must also use performance management practices that are driven by the natural performance cycles of the tasks that teams are doing. Various phased project reviews for new product development cycles provide a model for review processes that are timed by naturally occurring subcycles rather than according to a predetermined calendar cycle. Recent research on the performance cycles of project teams indicates that the halfway point of a project of fixed time duration is almost invariably a point of transition (Gersick, 1988) and that review of performance at that point will generally lead to helpful process changes. If ongoing performance management practices are in place that are multilevel and lateral in nature then these practices will allow transition points to occur in a productive way.

Movement of People

The movement of people through a succession of jobs or assignments is regulated by career ladders, succession planning, job posting, application, selection, and placement systems and practices. Much of this has not been done in a very planful way in many organizations (Von Glinow, 1988). The models that exist have generally been hierarchical both in terms of the locus of control for job movement and in terms of the desired direction of movement for careers. The assumed interest of the organization has been to place the most qualified people in positions of greater responsibility and authority, which has generally meant higher in the organization and with more authority over the allocation of resources and decision making. Status and reward systems have conformed to the notion that higher is more valuable, and consequently most employees find it in their self-interest (even if not their personal interest) to pursue "linear" (upward) careers.

This hierarchically oriented paradigm has always had its dysfunctional consequences, particularly for technologically oriented employees, who quite literally may have to leave their professional interests in order to make a career for themselves, generally in the ranks of management. They may not be suited for nor interested in managerial responsibilities (Bailyn, 1985; Miller, 1986). The success of dual and even triple ladders has been mixed. The concept of choice is viable, but the implementation often falls short of expectation because of failure to attach parity between the ladders either in pay or in recognition accorded to the importance of the positions (Raelin, 1987; Riggs, 1983).

As organizations become flatter, self-regulation becomes the norm, cross-functional teams and networks manage processes that cut across the organization, and temporary structures become the norm rather than the exception, a new model for people movement and for matching people and assignments becomes even more imperative.

The new model must address the needs of all levels of the system. Ongoing opportunities for learning, growth and development, leading to increased intrinsic motivation as well as opportunities for growth in salary and status are the interests of the individual. Company interest is in optimal deployment of its core skills and competencies to tasks critical to the execution of strategy, for continuous improvement in its performance capabilities, and for employees to develop the necessary base of expertise to perform effectively in self-regulating lateral structures. The need of teams is for adequate resources, and for continuity of membership as they go through the various stages of a project so that the project does not suffer from the loss of the cumulative learning that contributes to its effectiveness. In addition, the trends toward parallel processing, such as are inherent in simultaneous engineering and quality functional deployment models, push the organization in the direction of early and ongoing multifunctional dedication of resources. This need works against a purely commodity theory of technical hours whereby people with expensive skills are "carved up" and parcelled out without regard to team or project dynamics.

The high-technology company of the future may be shaped quite differently, with fewer levels, and smaller numbers in higher management. Operational management will be pushed down to teams and mini-businesses, and top management will be composed of people with interests, propensities, and expertise in strategic leadership. There will be a greater amount of upward influence on and participation in strategic decision making, so that the skills for executive leadership will be learned as individuals in various product-line, program- or product-oriented teams develop firsthand knowledge of the "bigger picture" (Resnick-West & Von Glinow, 1990) and manage their own adaptation to organizational strategy.

Lateral movement will become a more predominant career image based on various job rotational schema and knowledge-acquisition oriented sequences of developmental assignments. Choice is key, offering paths able to accomodate differences in orientation toward breadth or depth (the organization will continue to need in-depth technical experts), and acknowledging that orientations can change at different stages in a person's career. At the operational levels of the organization, managerial functions may become one of many skill blocks, permitting employees to move back and forth between technical and managerial responsibilities without threat of losing status or salary. This enables many more employees to be able both to continue to contribute technically and to develop an in-depth understanding of the business and managerial imperatives and of the big picture.

The team level of analysis will be the locus for a great deal of career definition and for the intersection of individual and team requirements. As cross-functional teams go through stages of a project, individuals are exposed to issues of many disciplines. Assignments within a team can be individually negotiated among team members to include interesting combinations of responsibilities and provide growth, development, and motivation. Careers will increasingly be thought of as a succession of position descriptions that are tailored to the current needs of the individual and the organization/team. Individuals will, in a sense, become "contractors" to the organization, taking responsibility of developing the skills and expertise to contribute as needed, and being recognized, rewarded, and given responsibility based on the extent to which they are able to demonstrate that value within the lateral network.

It is critical to this model that a placement and succession planning system be based on an information system that enables the organization to keep track of its skill pool and to assure deployment of critical skills to tasks of critical strategic impact. Likewise, the information system has to provide individuals with information concerning opportunities that will allow them to pursue their interests and to develop their skill base to ensure that they can continue to contribute. The team level in the organization is where execution occurs and it is where those two needs are fused. Teams will not only require the talent and expertise, but they will also provide the enforcement of performance norms and standards and will be the ultimate arbiters of an individual's value to the organization.

In the flatter organization, providing psychic rewards will be critical to the retention of key technical talent. Retention will depend on providing choice: multiple valued (recognized and rewarded) routes through the organization, operational autonomy at the team level, and the opportunity for access to strategic decision making and input to key decisions. In return, the individual will have to continually keep in mind his/her need to demonstrate value to the enterprise.

Reward Systems

As with other human resource practices the tradition in reward and compensation systems has strongly been to design these systems for individuals. Much of the reason lies in analytic approaches to organizing that lead to paying people based on the particular job they do. The problem is that this analytic logic works at odds with what high-technology organizations demand. Job based pay locks the organization into job structures that may become obsolete due to rapid change (Lawler, 1990). If jobs earn differential pay then a status hierarchy of jobs exists that becomes an incentive to move up the jobs in the hierarchy. When there is no pay incentive to move laterally among jobs of approximately equal worth, or into jobs that have been "valued" lower, it often becomes a stigma to do so. Since management jobs have come to be paid more, the incentive is for technical people to move out of the technical realm and into the managerial. Dual ladders have been created to short circuit this tendency by creating a technical ladder of jobs parallel with the managerial one. But dual ladders do not address the need for lateral movement among positions, rather they encourage individuals to stay in one discipline and get better and better (more and more specialized?) at doing it. If not combined with lateral "ladders," dual ladders of compensation reinforce the analytic and hierarchic approaches to rewards and compensation.

The high-technology organization of the future will require a pay system that reinforces its needs for cross-functional cooperation and understanding, and that reinforces the need for lateral cross-functional career paths. It must foster flexibility by encouraging employees to accept new roles, and must reinforce the need for individuals to develop new expertise as changes in organization and environment require. An approach with the potential to do these things is to base compensation on people's knowledge, skills and experience rather than on their job assignments (Lawler, 1990; Lawler & Ledford, 1985). Under this logic an individual acquiring more breadth of knowledge and skills would be seen as more valuable to the organization. For instance, an R&D engineer could assume a marketing role and as she added the knowledge and skills to perform in the marketing role to the engineering skills she already had, then she would receive increased base pay to represent the increase in skill breadth. This kind of knowledge-based pay system has been used very successfully in team based environments in manufacturing settings (Lawler & Ledford, 1985) and in some professional settings (Ledford, 1990). Recently some organizations have begun to institutionalize the approach company-wide, including managers (Ledford, 1990).

While a pay for knowledge plan can reinforce individuals gaining cross-functional understanding, it does not necessarily focus people on the task at hand. To focus people on the task at hand requires a reward system that is contingent on the accomplishment of that task. These kinds of rewards generally are considered to be "performance based" (Lawler, 1990).

Performance based rewards can be applied to any level: individual, team, and organization. They can focus people on the task associated with the level at which they are applied. In programs such as gain sharing and profit sharing, for example, the entire organization shares monetarily in the performance success of the organization. The formulas these programs use to measure performance and convert it into rewards are well-specified definitions of what the organization's performance is to be. Gainsharing and profit sharing programs that pay out to all the employees are excellent examples of human resource practices that are aimed at entire organizations. They reinforce the processes used to synthesize the parts of an organization into a whole.

Something like gainsharing could be contemplated at levels below the entire organization, such as teams. In order to do so, however, the team members must be interdependent with one another and the team must be responsible for some measurable overall output to which basic performance numbers and monetary worth can be attached. High-technology product development teams are potentially good examples, for instance. In defense or other contract oriented settings, the project members might share in the savings that result if the contract is delivered under cost or ahead of schedule. In commercial settings, project members might share in additional "profit" associated with a product by doing such things as getting it to market earlier, using less development costs, accurately designing to customer desires, and designing for low cost manufacturing.

Team oriented special awards programs can be used to reward teams that have done notable work. These reward programs differ from target or formula driven incentive approaches in that they are used after the fact to acknowledge excellent performance. They appear to have positive results in high-technology settings (Mohrman et al., 1990).

The nature of gainsharing and profit sharing are to focus attention to the performance of the whole system, but at the present time and in the short term. There are other, complementary, organizational level reward systems that also offer a piece of the action to employees but that focus on the longer run. These include various employee ownership approaches such as stock option plans. These also have been shown to be highly compatible with high-technology settings (Gomez-Mejia & Welbourne, 1990).

Organization level reward approaches, whether long or short term in focus, can successfully be used to foster organizational performance. But research shows that this only happens to the extent that the organization also contains an employee involvement culture and infrastructure that provide all employees with means to change the organization so that it can become more effective (Blinder, 1990). In other words, organization level reward systems only work to improve performance if they provide employees with ways of changing the organization to improve performance. This is an extremely important point. Organization level reward systems, therefore, also reinforce continuous organizational development and change and the mechanisms to accomplish them. This is highly compatible with the required dynamic nature of high technology organizations.

Both team level and organization level rewards support the need for lateral organization in high-technology settings. These reward systems highlight the superordinate goals that all the parts must share and therefore highlight the importance of the interdependent parts working together to achieve them. Conversely, performance based systems aimed at the parts (whether the parts are individuals or teams) can divert the focus of the parts away from the superordinate goals and onto those at their level. The trade-off is line of sight—the greater the level of aggregation that is rewarded, the less the component parts feel they can truly influence their outcomes. As we discussed in the section on performance management, individual performance must be defined in the context of group performance. This definition process helps the individual clarify the line of sight between individual performance and team outcomes.

It is risky to abandon individual reward and recognition altogether. Individual pay for performance practices are closely related to employee trust, satisfaction, and equity perceptions (Mohrman et al., 1990). In our strongly individualistic culture individual pay for performance may be necessary in organizations primarily to satisfy individually oriented perceptions of equity and of self-worth. Unfortunately, individual pay for performance systems are seldom seen as equitable because of the many factors in their administration that cause employees to perceive their merit pay as unfair (Lawler, 1990). In addition, fixed pool merit pay approaches typically set up zero sum games in which those who succeed do so at the expense of their coworkers (Deming, 1986a; Lawler, 1990). In high-technology organizations that we have studied, individual pay for performance has shown no impact on group performance and in some cases it is negative. Individual pay for performance had no effect on individual performance either (Mohrman et al., 1990). The design and implementation of individual pay approaches must be done in such a way as to foster the linkages of the individual to the larger system. Current practices do not work.

The logic of pay equity needs to be altered and expanded beyond that of individual pay based on job defined performance. Both knowledge-based pay systems and gainsharing are good alternatives. Knowledge-based pay systems have been shown to be perceived as even more equitable than individual pay for performance systems (Ledford, 1991). In addition, gainsharing has been shown to strengthen feelings of equity and satisfaction when used with knowledge-based pay (Mohrman, Ledford, & Demming, 1987). As performance begins to be carefully managed at multiple levels, the logic of equity may begin to include the expectation that a total reward package should address the performance of the multiple levels in which an individual is embedded. Because of the fast paced environment of high technology, and the speed at which organizations go through development stages and change their strategies, the best reward strategies allow ongoing changes in the design of rewards to reflect the current business direction of the organization (Schuster, 1984). The framework of the reward program must be flexible to accommodate diverse needs within the company, and must itself be tinkered with to ensure that the company has the reward tools it needs to reinforce its changing strategic and operating needs. If individuals are to feel rewarded by rather than victimized by such a dynamic approach to rewards, the development of the reward systems must be participative (Lawler, 1990), and occur in a context of rich, open information and understanding of the key performance values and business requirements (Schuster, 1984).

Human Resource Information Systems

Just as information technology is one of the forces driving the need and potential for lateral organizational forms, the development of integrated human resources information systems will be pivotal in supporting the human resources aspects of these organizations. The development of common corporate data bases and systems will enable the simultaneous management of human resources at all levels of aggregation. At the company level it will make possible a tracking and auditing of overall human resource capabilities—skills pools, experience pools, and demographics—as well as permit tracking and deployment of critical core competencies. Key executive placement and succession planning can be based on a solid foundation of longitudinal information rather than on limited data bases that describe the small pool that has emerged at relatively high organizational levels after years of opportunistic placement and selection based on limited information on the part of the hiring manager and the employee. Common data pools will also allow the tracking and targeting of development opportunities.

Common data systems are equally powerful in distributing information to empowered units for use in selection, team composition, and assembling the optimal mix of contributors from a skills and cost perspective. Opening up information about career opportunities to employees throughout the company enables more self direction and responsibility for careers.

Integrated human resources information systems allow the development and ease of administration of the kinds of multilevel pay and incentive packages that were discussed earlier and that result in a pay package tailored to the individual and based on a mixture of individual, project, and business performance. Likewise, performance feedback from a wide net of members of a person's role set can be collected and fed back for purposes of appraisal and performance planning and improvement.

Just as is true for other shared data systems, there will be many security and confidentiality issues that will have to be resolved. And yet the benefits of such a system in opening up information, supporting the ongoing, dynamic staffing needs of a dynamic series of performance units, and providing the simultaneous capability to individualize and at the same time to track and manage overall statistics and trends, will truly permit adaptive capability. The development of integrated and distributed human resource data systems is on the critical path to the establishment of the laterally integrated organization. Their use will be an underpinning of the emerging organizational forms.

The Challenge: Going Beyond Individualism and Control

Lateral integration challenges two of the fundamental values that characterize American organizations: individualism and hierarchical control. Individualism is a strong aspect of our society in general, and is instilled by a large number of societal factors, not the least of which are our educational institutions (Dertouzos et al., 1989). Technical training is no exception. The analytic work breakdown approach to organzation is quite compatible with the problem-solving paradigms taught in educational institutions (Adler, 1990). Engineers may arrive at their first jobs with "lone ranger" images of their future work life. They certainly arrive with expectations for clear specifications and operational autonomy (Von Glinow, 1988). The resolution of the discrepancy between this societal tendency and the requirements of high-technology organizations will be to find ways of managing that combine individualism and cooperation (Dertouzos et al., 1989), and that enable the management of human resources at multiple systemic levels.

Likewise, hierarchical control rather than self-control has been assumed as the underpinning of control in the essentially bureaucratic models of the organization. Entire organizational levels exist solely for this purpose—and have proliferated as organizations have tried to assure "manageable" spans of control. Although there is currently a societal awakening to the many costs of this form of control, the "controlling" interests of most corporations lie with those who have been taught their role is to control.

Like most large-scale change (Mohrman et al., 1989), the evolution to lateral, systemic approaches to management will be driven by environmental change—by the economic forces of the global competitive environment and by the potential that has been unleashed because of the advances in technology. Culture change it will be—but culture change it *must* be if our organizations are to be the leaders in the fast-paced, high-stakes arena of high technology.

CONCLUSION

This paper has addressed the need for increased reliance on lateral integration to address the extreme information processing needs of high technology organizations. Teams and networks will increasingly have to be managed. Whereas there is widespread recognition of the need for lateral integration, most organizations rely heavily on analytical and hierarchical organizing approaches. There is a need to develop organizational approaches that are more systemic in nature, and that build the ability to self design into the performing units of the organization. This will require a change in emphasis in human resource practices to acknowledge the different systemic levels of the organization (individuals, teams, networks, and whole organizations), to rely more heavily on lateral management, and to assume change rather than permanence.

ACKNOWLEDGMENT

An earlier version of this paper was presented at the Second International Conference on Managing the High Technology Firm, January 10-12, 1990, University of Colorado at Boulder.

REFERENCES

Ackoff, R.L. (1981). *Creating the corporate future.* New York: Wiley.

Adler, P. (1990). Managing high tech processes: The challenge of CAD/CAM. In M.A. Von Glinow & S.A. Mohrman (Eds.), *Managing complexity in high technology organizations* (pp. 188-215). New York: Oxford University Press.

Ancona, D.B. (1990). Top management teams: Preparing for the revolution. In J. Carroll (Ed.), *Applied social psychology and organizational settings* (pp. 99-128). Hillsdale, NJ: Lawrence Erlbaum Associates, Inc.

Anderson, C.S., & Kleingartner, A. (1987). Human resource management in high technology firms and the impact of professionalism. In A. Kleingartner & C.S. Anderson (Eds.), *Human resource management in high technology firms* (pp. 3-21). Lexington, MA: Lexington Books.

Applegate, L.M., Cash, J.I. Jr., & Mills, D.Q. (1988). Information technology and tomorrow's manager. *Harvard Business Review, 66,* 128-136.

Bailyn, L. (1985). Autonomy in the industrial R&D lab. *Human Resource Management, 24*(2), 129-146.

Blinder, A.S. (1990). *Paying for productivity.* Washington, DC: Brookings.

Birnbaum-More, P.H., & Nadler, G. (1989). New product development cycle time: A cross national study. Grant Proposal to the National Science Foundation.

Bower, J., & Hout, T. (1988). Fast-Cycle capability for competitive power. *Harvard Business Review, 66,* 110-118.

Crosby, P.B. (1979). *Quality is free.* New York: McGraw-Hill.

Deming, W.E. (1986a). *Out of the crisis.* Cambridge, MA: Productivity Press.

Deming, W.E. (1986b). *Quality, productivity, and competitive position.* Cambridge, MA: MIT Center for Advanced Engineering Study.

Dertouzos, M.L., Lester, R.K., & Solow, R.M. (1989). *Made in America: Regaining the productive edge.* Cambridge, MA: The MIT Press.

Dornbusch, S.M., & Scott, W.R. (1977). *Evaluation and the exercise of authority.* San Francisco, CA: Jossey-Bass.

Eisenhardt, K. (1989). Making fast strategic decisions in high velocity environments. *Academy of Management Journal,* 32, 543-576.

Eisenstat, R.A., & Cohen, S.G. (1990). Summary: Top management groups. In J.R. Hackman (Ed.), Groups that work (and those that don't): Creating conditions for effective teamwork (pp. 78-86). San Francisco,CA: Jossey-Bass.

Galbraith, J. (1973). *Designing complex organizations.* Reading, MA: Addison Wesley.

Galbraith, J.R. (1990). *Organization design: Past, present, and future.* Working Paper, Los Angeles, CA: The Center for Effective Organizations, University of Southern California.

Gersick, C.J.G. (1988). Time and transition in work teams: Toward a new model of group development. *Academy of Management Journal,* 31, 9-41.

Gomez-Mejia, L.R., & Welbourne, T.M. (1990). The role of compensation in the human resource management strategies of high technology firms. In M.A. Von Glinow & S.A. Mohrman (Eds.), *Managing Complexity in High Technology Organizations* (pp. 188-215). New York: Oxford University Press.

Gupta, A., & Wilemon, D. (1988). The credibility-cooperation connection at the R&D-Marketing interface. *Journal of Product Innovation Management,* 5, 20-31.

Haas, E. (1987). Breakthrough manufacturing. *Harvard Business Review, 63,* 75-82.

Johansen, R. (1988). *Groupware: Computer support for business teams.* New York: The Free Press.

Juran, J.M. (1964). *Managerial breakthrough.* New York: McGraw-Hill.

Juran, J.M. (1988). *Juran on planning for quality.* New York: The Free Press.

Juran, J.M., & Gryna, F.M. (1980). *Quality planning analysis.* New York: McGraw-Hill.

Kanter. R. (1983). *The change masters: Innovation for productivity in the American corporation.* New York: Simon & Shuster.

Lawler, E.E. III. (1990). *Strategic pay.* San Francisco, CA: Jossey-Bass.

Lawler, E.E. III, & Ledford, G.E., Jr. (1985). Skill-Based pay. *Personnel, 62*(9), 30-37.

Lawrence, P., & Lorsch, J. (1967). *Organizations and environments.* Homewood, IL: Irwin.

Ledford, G.E., Jr. (1990). The effectiveness of skill-based pay systems. *Perspectives in Total Compensation, 1*(1), 1-4.

Ledford, G.E., Jr. (1991). Three case studies of skill-based pay: An overview. *Compensation and Benefits Review, 23*(2), 11-23.

Miller, D.B. (1986. *Managing professionals in research and development.* San Francisco, CA: Jossey-Bass.

Mohrman, A.M., Jr., Ledford, G.E., Jr., & Demming, S. (1987). *Gainsharing: Congruence with high involvement organization design.* Working paper, Center for Effective Organizations, University of Southern California, Los Angeles, California.

Mohrman, A.M. Jr., Mohrman, S.A., Ledford, G.E., Jr., Cummings, T.G., Lawler, E.E. III, & Associates (1989). *Large scale organizational change.* San Francisco, CA: Jossey-Bass.

Mohrman, A.M., Mohrman, S.A., & Worley, C. (1990). High technology performance management. In M.A. Von Glinow & S.A. Mohrman (Eds.), *Managing Complexity in High Technology Organizations* (pp. 216-236). New York: Oxford Press.

Mohrman, S.A., & Cummings, T.G. (1989). *Self-Designing organizations: Learning how to create high performance.* Reading, MA: Addison-Wesley.

Pinchot, G. (1985). *Intrapreneurship.* New York: Harper & Row.

Raelin, J. (1987). Two-Track plans for one track careers. *Personnel Journal, 66*(1), 96-101.

Resnick-West, S., & Von Glinow, M. A. (1990). Beyond the clash: Managing high technology professionals. In M.A. Von Glinow & S.A. Mohrman (Eds.), *Managing complexity in high technology organizations* (pp. 237-254). New York: Oxford University Press.

Riggs, H. E. (1983). *Managing high-technology companies.* Belmont, CA: Lifetime Learning Publications.

Savage, C.M. (1990). *5th generation management: Integrating enterprises through human networking.* Maynard, MA.: Digital Press.

Schoonhoven, C.B., & Jelinek, M. (1990). Dynamic tension in innovative, high technology firms: Managing rapid technological change through organizational structure. In M.A. Von Glinow & S.A. Mohrman (Eds.), *Managing complexity in high technology organizations* (pp. 90-118). New York: Oxford Press.

Schuster, J. (1984). *Management compensation in high technology companies.* Lexington, MA: Lexington Books.

Souder, W.E. (1988). Managing relations between R&D and marketing in new product development projects. *Journal of Product Innovation and Management, 5*(1), 6-19.

Stalk, G., Jr. (1988). Time: The next source of competitive advantage. *Harvard Business Review, 66,* 41-51.

Takeuchi, H., & Nonaka, I. (1986). The new product development game. *Harvard Business Review, 64,* 137-146.

Thompson, J.D. (1967). *Organizations in action.* New York: McGraw-Hill.

Von Glinow, M.A. (1988). *The new professionals.* Cambridge, MA: Ballinger.

COMPETITIVE COMPENSATION STRATEGIES IN HIGH-TECH FIRMS

Marick F. Masters, George C. Tokesky, William S. Brown, Robert Atkin, and Gerald Schoenfeld

INTRODUCTION

The so-called "high-tech" firms are, interestingly, a vital part of the United States economy. As traditional industries suffer under foreign competition and are less able to provide growth and employment, it falls to new high-tech firms to be a source of vitality for the economy. Because of the highly competitive environment and rapid growth, the management of such firms, and the people therein, is a major challenge; since high-tech firms thrive on entrepreneurship and creativity, their success can be dependent on the skills and daring of a few. The need to attract and retain the right people is critical. One of the key elements of people management in high-tech firms is the compensation system, and the design of compensation programs must be in keeping with the nature of the business and the type of people who are attracted to this kind of career.

Human Resource Strategy in High Technology.
Advances in Global High-Technology Management, Volume 1, pages 59-87.

ISBN: 1-55938-489-1

High-tech firms have certain features in common that help define them as such. These are (1) a product that is at the leading edge of technology, (2) a high priority on research and development, (3) innovations that are frequent and numerous, (4) a high rate of turnover of R&D personnel, (5) a certain geographic concentration, and (6) high mortality rates (Balkin & Gomez-Mejia, 1984). Because of these unique characteristics, compensation systems may need to be different from traditional firms (Schuster, 1985). Generally, high-tech firms exhibit a greater reliance on incentive compensation, particularly while they are going through the growth stage of the product cycle (Balkin & Gomez-Mejia, 1984). It is in this context that we will consider the link between compensation strategies, including the fashionable "pay for performance" theme, and the nature of high-tech firms.

It is important that pay systems are designed to attract, retain, and motivate the right people. Even in the absence of a feeling that pay is, in itself, a motivator (e.g., Herzberg, Mausner, & Snyderman, 1959), pay systems surely must, at least, avoid disrupting the dynamics that attract, retain, and motivate (Lawler, 1983). Given the complexity of those dynamics—of the intrinsic and extrinsic factors that combine in a synergistic way to satisfy and motivate—it may not be possible to isolate the influence of the pay system on the success of the firm. Furthermore, considering the high mortality of high-tech firms and rapid turnover of key R&D personnel, identified above as characteristics of those firms, management may not have the luxury of experimenting with the pay system; instead, there is a real need to "get it right the first time." These two factors argue for the need for sound theory in designing pay systems for high-tech. Part of the purpose of this paper is, therefore, to examine the theoretical bases of compensation strategies in high-tech firms.

COMPENSATION STRATEGIES AND MOTIVATIONAL MODELS

Theories of motivation are built upon the theorist's paradigms of the nature of human beings and of the human condition. The idea of human beings as self-seeking, practical creatures led rise to a hedonistic utilitarian argument for motivation. In this view, people naturally seek to increase happiness and diminish misery. Hull (1943, 1951, 1952) saw behavior as a function of drive-response and strength of habits. The activation of these drives occurs when there is a psychological or physiological deprivation (need) which results in behaviors to alleviate that deprivation. Deprivation, therefore, is claimed to be the basis of motivation (Williams, 1982). Langer (1937) developed a theory of unconscious needs satisfaction and added the concept of a priority of needs. Maslow (1943, 1954) followed with his now-famous hierarchy of needs theory. To the extent that compensation is seen as primarily servicing lower needs (i.e., physiological and safety needs), one might theorize that compensation is limited in its potential as a motivator, and that other motivators are required to service higher needs. However, one should not overlook the possible symbolic value of compensation. First, cash rewards can be used to acquire materialistic symbols of achievement. Second, compensation may be intrinsically valued by the organization member as a tangible measure of self-worth. This may be particularly relevant to high-tech firms, wherein entrepreneurial egos may be serviced with tangible rewards.

The idea of the dual value of rewards (to fulfill lower needs and higher needs) may be more easily seen as relevant if the hierarchy of needs is collapsed into a two-level rather than five-level scheme (with physiological and safety need comprising the first level, and belongingness, self-esteem, and self-actualization combined as the higher level), and, if different level needs are seen as occurring simultaneously rather than in priority order (Hall & Nougaim, 1968; Lawler & Suttle, 1972; Wahba & Birdwell, 1976). A study by Dacher and Hulin (1969) further suggested that satisfaction of a need does not extinguish motivation for that need, but rather, may enhance the drive for continuing satisfaction. Compensation practices obviously assume that the satisfaction drive is continuing (i.e., compensation structures are periodically revised).

Herzberg et al. (1959) also proposed a theory based on the duality of needs. In Herzberg's scheme, all work-related factors can be divided into two categories, motivation factors (satisfiers) and maintenance or hygiene factors (possible dissatisfiers). The factors considered to be motivators were related to job content (intrinsic), whereas the factors considered to be hygiene factors were related to job context (extrinsic). Pay, to Herzberg, "has more potency as a job dissatisfier than a job satisfier." However, Lawler (1971) recognized that, while inequitably low pay is a dissatisfier, pay can also be used to satisfy esteem and recognition needs, and, as such, can be a motivator, in addition to being a hygiene factor.

Somewhat related to the notion that individuals are hedonistic and utilitarian, is the general class of motivational theories that might be termed economic and/or bargaining theories. The premise is that people are economic, rational beings, who will naturally seek to maximize utility (e.g., Nash, 1950; Young, 1975). The promise of gain leads to willingness to participate in the organization's efforts. Barnard (1938) suggested a simple equation by which "the net satisfactions which induce a man to contribute his efforts to an organization result from the positive advantages as against the disadvantages." Among those "positive advantages" were monetary rewards. March and Simon (1958) further supposed a utility value for "each component of a set of inducements." In some cases, possibly high-tech firms in particular, contingency pay arrangements may be an agreement to reward specific achievements. Shackle (1949) added the idea of probability of an outcome, to the utility of an outcome, which brings bargaining theory closer to expectancy theory (discussed below).

Etzioni (1961) provided an organizational sociology perspective, which appears to have relevance to compensation strategies. He examined behavior compliance patterns within the organization, in the context of power applied. Three types of power considered were coercive, remunerative, and normative (the allocation and manipulation of symbolic rewards). Compensation could be used to exercise control, both as an obvious application of remunerative power, and also as a form of normative power, if one accepts the notion of the symbolic value of compensation. This also corresponds to the two levels of human needs, as discussed above.

Agency theory has origins both in economic theory and in power theory (Eisenhardt, 1985, 1988). Agency theory deals with the relationship of the "principal" (e.g., owner) and "agent" (e.g., manager, employee). The principal provides rewards for the agent's performance in accordance with explicit or implicit contractual arrangements. Agency theory looks at two general categories of performance measurement: (1) outcomes and (2) behaviors. To the extent that the principal and agent have different goals, and information about the agent's behavior is limited or expensive, performance may be defined as an outcome (e.g., profits or return on equity). If the goals of the principal and agent are less disparate (especially if the agent shares in ownership) or if the principal's information about the internal workings of the organization is good, performance can be defined in terms of behavior (Eisenhardt, 1989). Behavioral definitions of performance may be preferable if outcomes are difficult to measure, or if the principal has more of a longer-term perspective rather than only short-term goals. In either case, rewards to the agent can be the link between the requirements of the principal and the performance of the agent (Oviatt, 1988).

Agency theory has provided some research on compensation relative to workplace motivation (Conlon & Parks, 1988; Eisenhardt, 1985, 1988), but there has been particular interest in the specific area of top executive compensation (Pearce, Stevenson & Perry, 1985; Ungson & Stears, 1984). The total compensation package, including cash rewards, indirect economic rewards (benefits, ownership), and nonmonetary rewards (e.g., time off, status symbols) must be appropriate relative to the goals of the owners and the goals, needs, and desires of the agent. To the extent that the needs of the agent are deeply-rooted, rewards have significant motivational potential (Gomez-Mejia & Welbourne, 1988: Milkovich, 1988). It is important to recognize differences in needs among different types of employees (Lawler, 1971). Furthermore, perception of compensation and compensation equity varies and can be crucial to motivation (Fin & Lee, 1972, Lawler, 1983; Wernimont & Fitzpatrick, 1972).

Lawler (1976, 1983) suggested the concept of "organizational congruence," in which the design of pay system is dependent on certain organizational factors, such as the technology of the business, the organization's size, the human relations climate, and the task characteristics. To the extent that high-tech firms are different in these aspects from traditional firms, congruency theory may be particularly relevant in compensation strategy.

Learning theory considers the way in which past behavioral outcomes (rewards and punishment) influence future behaviors (effort and performance). Positive behavior is shaped and reinforced by the delivery of rewards. Negative behavior is discouraged by withholding rewards, or by punishment (Skinner, 1953). In order for learning or shaping to take place, the individual must be able to clearly associate the reward with the behavior. Good communication and performance appraisal would, therefore, be a requirement. Behavior modification has been proposed and used in organizations with reported success (Schneier, 1974), although many managers may be uncomfortable with the suggestion that behavior modification is manipulative and intrusive. Variable schedule rewards are thought to be particularly potent (Berger, Cummings, & Heneman, 1975; Saari & Latham, 1982; Yukl, Wexley, & Seymore, 1972), so one might suppose that pay systems should allow for the delivery of rewards on other than a traditional 12-month cycle.

Expectancy theory (Vroom, 1964) is often used to examine the pay/motivation or pay/performance relationship (Heneman & Schwab, 1972; Lawler, 1971, 1983, Lawler & Suttle, 1973). It has some common features with learning theory and utility theory. If anticipated rewards are positive (valence), significant, and valuable (instrumentality), and have a high probability of being obtained (expectancy), the employee will presumably be motivated to make an effort to perform (e.g., Ilgen, Nebeker, & Prichard, 1981). Future expectancies are influenced by past experience. If compensation is to be a motivator, the value of the reward must be significant (Campbell, Dunnette, Lawler, & Weick, 1970; Lawler, 1971, 1983; Porter & Lawler, 1965;). From expectancy theory, one might predict that minor gradients in merit pay increases or bonus awards would not be very motivational.

Equity theory (Adams, 1963, 1965) is particularly useful as a theoretical basis for pay-for-performance systems. It suggests that people are motivated by the perceived fairness of rewards. Herzberg et al. (1959) observed that feelings of unfairness were the most frequently occurring source of job dissatisfaction. Adams (1963, 1965) saw the employer/employee relationship as an exchange equation, in which the employee gives effort (inputs) in return for some reward (outcomes). The employee compares the input-outcome relationship to that of a self-defined reference group. When the employee's input-outcome ratio does not compare favorably to the reference group, there is a perception of inequity; the inequity can be a feeling of being either under-rewarded or over-rewarded. Perceived inequity results in either anger, in the case of under- reward (Garland 1973; Kessler & Wiener, 1972) or guilt, in the case of over-reward (Andrews, 1967; Zadnek & Smith, 1968). Adams was influenced by the earlier work on cognitive dissonance, by Festinger (1957), and distributive justice, by Homans (1961).

There are three general measures of equity: external equity, internal equity, and individual equity (Lawler, 1983). Dyer and Theriault (1976) provided a noteworthy variant to the model, in adding perceived adequacy of pay administration. External equity looks at the market value of a job, while internal equity considers the relative value of a job to the firm. Individual equity compares peers, usually with respect to personal performance, experience, and so forth. People may use a number of referents to determine the perceived fairness of rewards gained, including superiors, subordinates, peers inside the firm, and peers outside the firm. A problem is that perceptions of equity may be seriously incorrect due to inaccurate information about those referents, or inappropriate selection of referents (Lawler, 1972; Patchen, 1961). This is especially true when the pay system is kept secret (Lawler, 1983). Even with disclosed pay systems, the information provided by management to the employees may not be believed. Equity theory is focused more on the reinforcement end of the effort/performance/reward/effort cycle (i.e., equitable rewards should lead to future effort). Two conditions are required, however: (1) a clear perception must exist of the nature and volume of inputs required to gain a reward, and (2) a clear perception must exist with respect to one's position relative to the reference group (Cook, 1975).

McGregor (1960) offered a view of two styles of management, "Theory X" and "Theory Y." The "Theory X" manager has a negative view of human beings in the workplace (i.e., that people dislike work and have to be controlled and coerced). The role of rewards systems, given this assumption, would be to threaten withholding of rewards in order to get workers to put forth adequate effort. The "Theory Y" manager assumes that effort in work is a natural inclination of people, and that people will take responsibility if encouraged to do so. Rewards in such context take on more the character of reinforcement. McGregor also suggested that management could profit by gaining employee participation in the direction of the organization. In this vein, Lawler (1983) proposed that worker satisfaction with rewards could be increased by participation in the design of rewards systems, especially in the design of incentive plans. High-tech firms, where incentives are often used, might consider this option.

PAY FOR PERFORMANCE

Given the leading edge, innovative, entrepreneurial character of high-tech firms, and noting a high rate of turnover of R&D personnel (Balkin & Gomez-Mejia, 1984), pay for performance systems may be of particular interest. Pay for performance is only one of several models for delivering compensation to employees, but, for the last several years, it has been one of the most widely discussed models, both in the academic and practitioner literature (e.g., Evans, 1970; Hills, 1979; Kopelman & Reinharth, 1982; Lawler, 1971, 1983; Meyer, 1975; Schwab & Dyer, 1973). In the practitioner literature, it is often extolled as being able to deliver "retention of high performers; turnover of poor performers; higher individual performance; employee trust . . . ; higher employee morale" (Bates, 1981).

Pay for performance is based on the premise that there should be no "automatic" (i.e., general or cost of living) increases, but, rather, that all pay increase decisions should be based on the employee's personal efforts and accomplishments (Bates, 1981). In such a scheme, pay increases can vary widely, and it is not necessary to give an increase at all, if none is perceived, by management, to be "deserved."

In spite of the increasing popularity of pay for performance, there are many objections to this model. One objection is that firms generally lack good performance measures, and, therefore, cannot truly pay for performance (Kearney, 1985). Another is that pay for performance may set up divisive competition within the organization, and may be counterproductive with respect to team building. A third is that some compensation to offset inflation is required for all employees. There are, therefore, alternative models for pay increases, and, even where organizations profess to have a pay for performance system, actual practices may draw upon other models, as well.

A Pay for Performance Model

Figure 1 shows a model for pay for performance. The individual measures himself or herself in terms of effort expended. Effort may or may not translate into performance, depending on the intensity of the effort, certain job and individual variables, and exogenous circumstances. Job variables include opportunities that the job presents for the individual to have an impact on the organization, the control the individual has over his or her situation, and the degree of dependence on others. Personal variables include skill, knowledge, and an understanding of what is required to bring results, particularly the extent to which the individual's skills fit the requirements of the job. Exogenous circumstances might include competition and the market's relative vitality.

Assuming that performance differences can be measured, or, at least, perceived, the question remains whether or not the organization is willing to pay for performance. Willingness has at least three aspects: (1) the organization's fiscal ability and willingness to spend money for performance increases, (2) the organization's commitment to a pay for performance ethic, and (3) the individual manager's willingness to recognize performance differentials. Even if the organization is willing and able to bear the expense and has a meaningful pay for performance program (i.e., a pay policy that

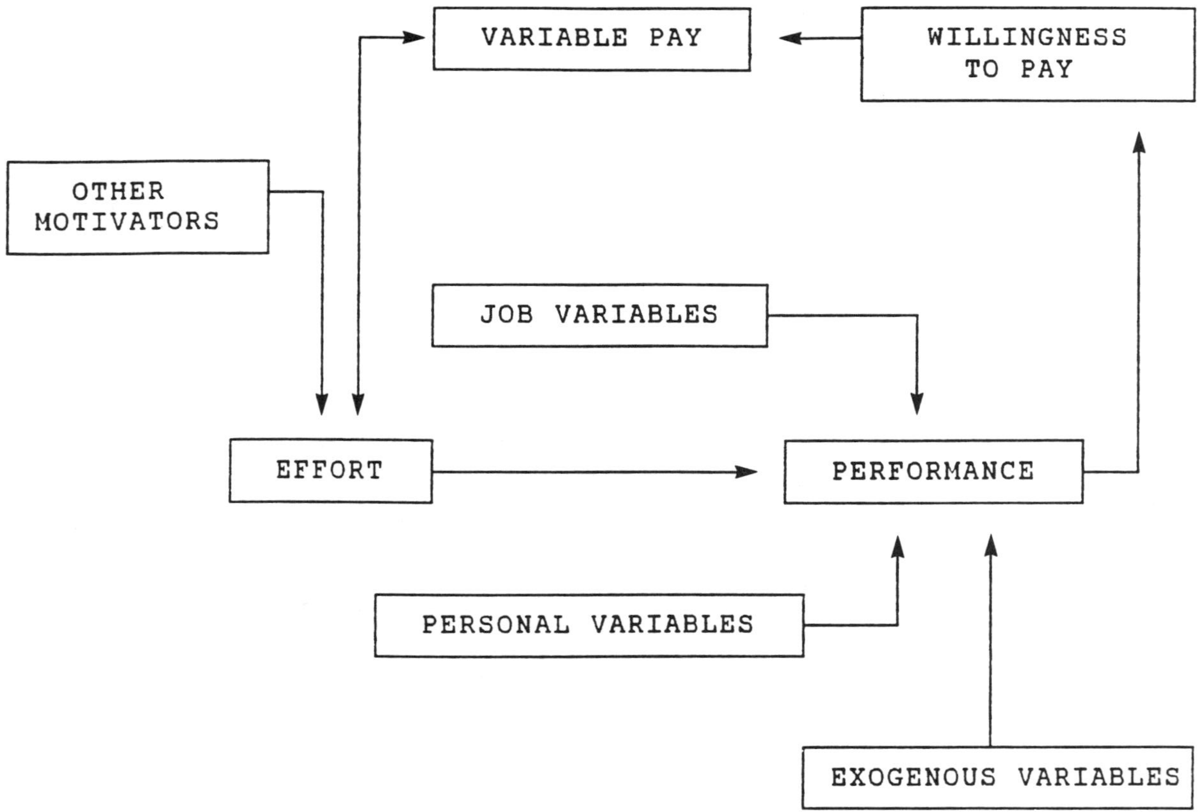

Figure 1. A Model for Pay for Performance

permits a significant range of increases), individual managers may not be willing to make the "hard" decisions required to differentiate between levels of performance. Many managers will feel compelled to dispense some minimum increase to all subordinates, and others will feel quite uncomfortable about awarding any very large increases; the result is differentials of a percent or two, which is hardly very exciting to the employee.

If the firm is into a pay for performance mode, performance will result in rewards, in the form of compensation. In turn, positive rewards should stimulate future positive behavior, or the absence of positive rewards should send a message that a change in behavior is required. There are other factors, other than compensation, that will impact behavior, including other rewards and negative factors, which may be both intrinsic and extrinsic to the job (Daniel & Esser, 1980; Mawhinney, 1979; O'Reilly & Caldwell, 1980). Intrinsic factors could include the challenge of the job itself, the amount of self-esteem associated with success relative to the challenge, the interpersonal relationships that the job generates, job stress, conflict between one's work life and personal life, congruity with one's social and personal values, physical working conditions, and so forth. Extrinsic factors include the social recognition and prestige associated with the job, positive feedback from superiors and peers, and noncash rewards, such as benefits and perquisites, and so forth.

An additional confounding factor with pay for performance systems is that negative messages sent through the rewards system are often not understood or believed (Russell, Gallagher, & Tokesky, 1986). Attribution theory (Downey, Chacko, & McElroy, 1979; Staw, 1975) suggests that people tend to attribute their successes to internal factors, such as their own ability, but blame their failures on external factors, such as the "system," the inability of one's superiors, lack of cooperation from one's peers, market circumstances, bad luck, and so forth.

Types of Pay for Performance Systems

Pay for performance could be delivered in three or possibly four different formats. Three clearly different methods are: (1) base salary merit pay systems, (2) contingency pay programs with a relatively constant, but adequate, base salary (with the variable component usually delivered in the form of some sort of bonus), and (3) contingency pay programs in which the individual's standard of living is at risk because of a relatively low base salary (e.g., field sales additional compensation systems). A possible fourth method of delivery could be specialized compensation program designs for executives and key individuals. Executive compensation tends to be a bit different because of the level of the rewards and the corresponding level of expectations, which somewhat creates a special subculture in the organization. The other way in which executive compensation differs is the complex manner in which rewards are delivered (e.g., base salary, short-term bonus awards, stock options, stock grants, stock appreciation rights, phantom stocks, performance share plans, deferred compensation, special benefits and retirement programs, perquisites, "golden parachutes," and so forth).

Another part of the design of a pay for performance system is how one chooses to define "performance" (Kearney, 1985). Three different, but equally reasonable, definitions might be (1) individual performance, (2) small group performance, and (3) organization performance. Furthermore, individual performance could be defined in terms of quantifiable outcomes (e.g., "Management by Objectives"; Drucker, 1954; McGregor, 1960; Odiorne, 1964) or in more subjective and perceptual terms (performance appraisal systems). If it can be argued that there is a relationship between performance (or, at least, effort) and satisfaction, there should be some consideration of how employee satisfaction enters into the design of systems.

Alternative Models

One alternative model is pay for job holding. In this model, pay is based on the value of the job to the organization or the importance of the individual to the organization, rather than the individual performance of the job incumbent (Pfeiffer & Davis-Blake, 1987). Most hourly-paid production workers are paid a given rate for a given job. Management jobs are often sorted by job evaluation (e.g., Belcher, 1974; Lawler, 1983), with those jobs having the highest value (i.e., scoring the most points in evaluation) receiving the highest pay and largest absolute dollar pay increases.

Another pay increase model is pay for membership, in which, the primary objective of the pay system is to simply attract and retain employees. This is similar to pay for job holding, but the emphasis is more on the individual than the job. This is something of what March and Simon (1958) referred to as "motivation to participate." Managers who are concerned that pay for performance sets up unhealthy competition inside the organization may prefer this model. Increases are generally an equal percentage to all employees, based on organization performance, movement of the employment market, or the cost of living. Cost of living can be an intentional or unintentional factor in so-called "merit pay" systems, as well (Dyer, Schwab, & Teriault, 1976), as can top-down budget constraints (Fossum & Fitch, 1985; Foster & Lynn, 1978).

Still another model, and one that has received some recent attention in the practitioner literature, is pay for skill or pay for knowledge. Employees are paid according to their personal qualifications or for their educational credentials (Lawler, 1983; Lawler & Ledford, 1985; Luthans & Fox, 1989). Pay systems that pay the individual rather than the job have been common in professional organizations, but have more recently been applied to lower-level production workers and service workers. On the shop floor or in a service facility, this may translate into the number of jobs for which one is qualified, or the number of tasks one is trained to perform (Gupta et al., 1986).

A Pay for Contribution Model

Job-evaluation based systems can be combined with the pay-for-performance ethic in what we might refer to as a "pay for contribution" model. In such a scheme, there are two elements to compensation: (1) the value of the job, and (2) the personal performance of the individual, within the scope of that job. Although the pay-for-performance element is often extolled, in truth, the component of pay determined by the job size is likely to outweigh the performance component. Figure 2 offers a model for pay for contribution.

When management, or the compensation practitioner, or both, are determining an appropriate level of rewards for contribution, consideration is given to internal equity, external equity, and individual equity (Lawler, 1983), as described above in the discussion of equity theory. Job analysis and job evaluation provides some measure of the relative value of the job to the organization, often in a point-score format. Job evaluation systems are valid to the extent that they can compare unlike jobs and differentiate between jobs that appear to be alike, but really are not (e.g., Belcher, 1974; Cascio 1982; Treiman, 1979). In this fashion, internal equity is dealt with. Compensation practitioners will turn to outside wage and salary surveys to determine the value of jobs on the employment market, which is external equity. Job evaluation and market information can be combined mathematically, using regression analysis, to determine a pay practice line for the organization. Generally, external equity will determine the overall slope and intercept of the line, while internal equity will prevail in pricing any given job. The pay policy line will establish a central value, or "midpoint," for every job, which is the major component of pay (e.g., Belcher, 1974; Cascio, 1982). Hopefully, this information will be shared with the employee, to establish a reference group for his or her personal sense of equity.

Some measure of individual equity among employees holding jobs of similar value will be determined by the pay for performance component of the compensation system. Variation in personal performance will result in variance of pay around the midpoint of the job. The individual performance component is a function of skill and personal effort. The employee has a perception of his or her own skill and effort, which may be reinforced or corrected through performance appraisal and employee development programs (e.g., Cascio, 1982; Landy & Farr, 1983).

Therefore, the level of responsibility of the job (or the "worth" or "value" of the job) and the individual's performance within the parameters of the job, combine as a measure of contribution to the organization. To the extent that variation in contribution will result invariable pay that is equitable, the

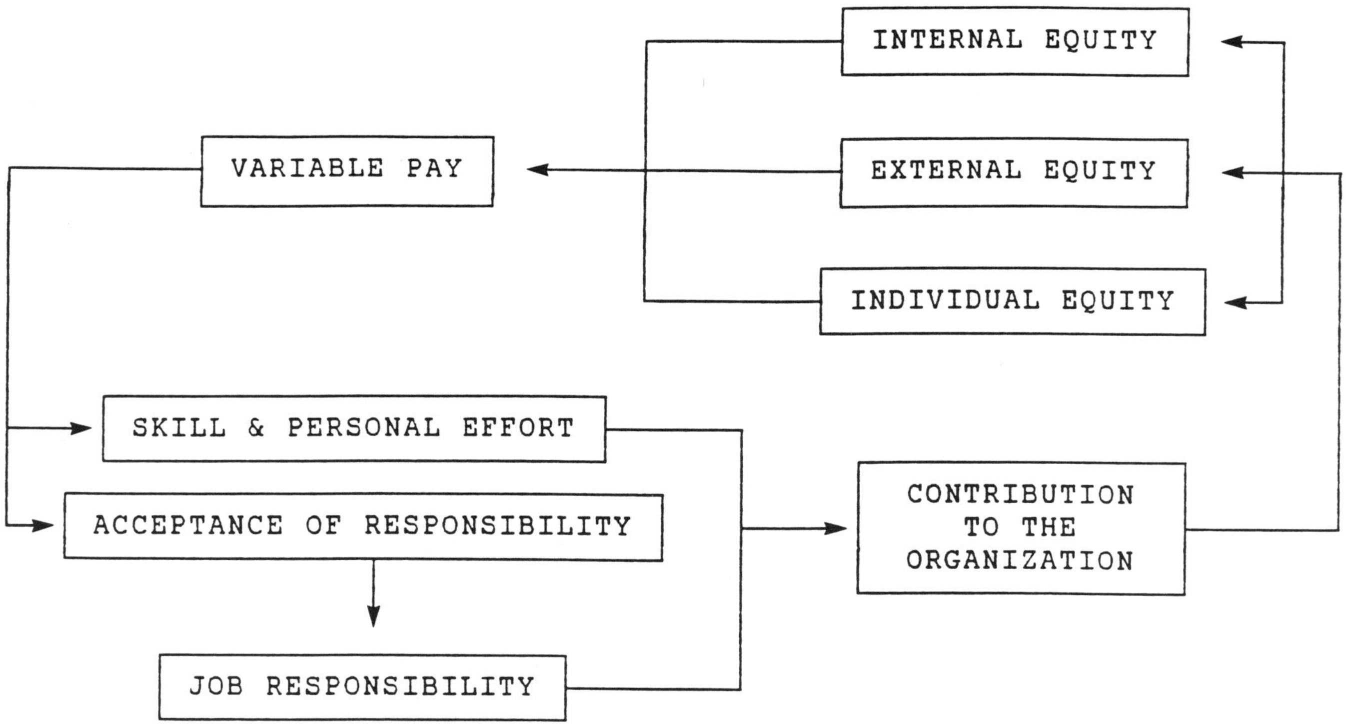

Figure 2. A Model for Pay for Contribution

employee will presumably be motivated to increase skill and effort, and to accept responsibility.

MOTIVATIONAL CONSIDERATIONS IN COMPENSATION SYSTEMS DESIGN

It would probably not be terribly astounding to suggest that workplace motivation is a complex algorithm, made up of a number of job variables, personal variables, social and cultural values, and so forth; compensation certainly may be an important factor, but we have to recognize the fact that it is only one of many influences on the individual employee. Furthermore, we might suggest that the motivation algorithm is different for different types of employees (Savery, 1988) and different for each individual. This is likely one reason why empirical work in compensation and satisfaction is so frustrating for the researcher; in fact, the pure empirical evidence supporting pay and performance or pay and satisfaction is rather thin, at best (Ehrenberg, 1990; Ehrenberg & Milkovich, 1987). Nevertheless, one has to try to make some sense of the whole business by assuming that certain generalizations are valid and useful.

There is some suggestion that pay has both economic and symbolic value (Lawler, 1971), and that pay, like all other satisfaction and motivation variables, may mean different things to different people. Krefting and Mahoney (1977), in searching for some definition of what constitutes a meaningful pay increase, tried dividing their subjects into those who thought of themselves as being motivated by money and those who thought of themselves as being motivated by recognition. Not surprisingly, those who identified themselves as being money-motivated were more influenced by consumption variables (cost of living, living standard, and income) relative to a meaningful pay increase. Those who said they were recognition-motivated were more influenced by perception of a "normal" pay increase—based on pay history, organization pay practices, and perception of the pay of others (i.e., by perception of equity). However, in the case of both groups, the amount of variance in predicting a meaningful pay increase was reduced when combinations of factors were considered. Which is to say, the money-motivated group was also influenced by perception of a normal pay increase (equity) and the recognition-motivated group was also influenced by consumption variables. In other words, those who were mainly money-motivated were also partly recognition-motivated, and those who were mainly recognition-motivated were also somewhat money-motivated. Interestingly enough, in Krefting and Mahoney's research, pay level was not a significant factor.

Those managers, practitioners, and academics who still doubt the effectiveness of compensation as a motivator, either for the consumption value of money or the recognition value, should, at the very least, be sensitive to the potency of pay (or the lack of pay) as a demotivator (Herzberg et al., 1959; Lawler, 1971), and compensation systems should be designed with adequate respect to the damage that can be done. Even if an individual is not particularly motivated by money, a perceived inequity in rewards is likely to quickly lead to confusion, disappointment, anger, and so forth (Adams, 1963, 1965). In all such cases, time and energy expended on resentment, worry, discussion with co-workers, and so forth will detract from time and energy available for performing the job. In extreme cases, the motivation to participate (March & Simon, 1958) is lost, resulting either in turnover or, worse yet, passive compliance with minimal work standards, while not turning over.

The opposite design error made by line managers and compensation practitioners is that behaviors and performance can be driven by pay. Other factors, such as leadership, job design, working conditions, participative management styles, and so forth, may be, in their minds, conveniently ignored. High pay can serve to attract and retain, but this can be a false economy, if job performance does not follow. In some cases, high pay can serve to retain employees who are otherwise dissatisfied or are in a poor job fit, simply because the alternative job opportunities do not provide equal economic rewards; this can be particularly true once the employee's life style and economic needs are pretty well established.

The compensation system design question is, then, not simply whether or not pay is a motivator (or demotivator), but more precisely, what is the role of compensation in the motivation equation? At one extreme of a continuum of theories, is the idea of a pure economic exchange of labor for pay; this may be more-or-less appropriate in situations where the job has little intrinsic interest and the worker has little else to offer other than labor. The opposite situation might be in humanistic professions, such as social work, or in volunteer work, where the self-gratification derived from the job itself presumably outweighs any monetary reward. However, in explaining the role of compensation in a whole range of other jobs and professions, including high-tech firms, we might suggest an alternative perspective, which will disagree with both the idea that pay is a driving force in the employer-employee contract and the idea that pay is not a motivator, at all. The notion we offer is that pay is very much a motivator, both for economic and recognition reasons, but that it is more of a reinforcing motivator than a driving motivator. In such a view, we would say that pay cannot be a substitute for human resources factors such as leadership, social rewards derived from membership in the work group, the intrinsic interest of the job (job design), participation in the decision processes and control over one's work situation, and so forth, but, that pay can, and should, significantly reinforce the culture and work climate of the organization.

Equity theory, would in fact, suggest an a posteriori satisfaction or dissatisfaction with rewards, based on comparison of one's actual reward outcomes to a reference group; the reference is probably not fixed, but, rather, "floats" over time. While expectancy theory has a more a priori flavor, it is likely that the expectancy equation changes as experience alters one's expectation of the probability and value of rewards. Looking at multiple cycles of expectations and the resulting behavior brings expectancy theory closer to learning theory, in which positive and negative reinforcements are critical to shaping behavior.

Which begs the question, Is pay satisfaction somewhat of a learned condition? The valuation of rewards is, in fact, a cognitive process, and perception of the value of an outcome is often adjusted with new information (Festinger, 1957; Gould, 1979; March & Simon, 1958). In Krefting and Mahoney's (1977) study, discussed above, the perception of a "normal" pay increase based on past experience, perception of the experiences of others, and knowledge of the organization's pay policy. Russell et al. (1986), found not much difference in pay satisfaction when the firm being studied went from a 7% pay increase program to a 4% pay increase program; an employee receiving a 6% increase under the 7% program would likely be disappointed, while an employee receiving a 5% increase under the 4% program might be quite pleased, unless some outside source, such as the industry average, provided the employee with significantly different information. If satisfaction is somewhat learned, it might take more than one reward cycle to determine whether or not a compensation program design was effective. This may be another confounding variable in the empirical study of pay and satisfaction or pay and performance, since research in this area is more typically cross-sectional than longitudinal in design (Russell et al., 1986). What this suggests is that it may be necessary to establish a pay system based on theory, beliefs, and culture, then hold to it for a time before giving it up as a bad idea.

We should be sensitive to the idea that there are two levels of human needs to be considered in compensation design. The first includes physiological and safety needs, which can be addressed by assuring that pay levels are adequate to provide security to employees. However, higher-level needs can be addressed by compensation systems as well, if one assumes that compensation has direct and indirect symbolic meaning, in addition to economic meaning (Lawler, 1971). Therefore, it is not reasonable to suppose that only intrinsic work values can provide the motivation that comes from powerful higher level needs. Extrinsic rewards can also provide partial satisfaction and reinforcement of these needs. At the very least, compensation strategies should not be contrary to the higher level needs of the employees. From a power perspective, control of the sources of needs satisfaction is a significant management tool, regardless of whether that power is used in a positive or threatening fashion.

Where the top management of a high-tech enterprise is not the same person or persons as the owner(s), agency theory can provide some understanding of the use of rewards to tie together the interests of these two groups (principal and agent). If the owner is more remote from the operation of the firm, rewards should be based on objective measures of performance (i.e., outcomes, such as profits or return on equity). If the owner is close to the operation of the firm and to management, job performance can be defined in terms of behaviors, and rewarded accordingly. The advantage of behavior-based rewards is that the owner can reward the long-term as well as the short-term, in that objective measures tend to be more short-term, but behavior measures can be both long-term and short-term.

Pay for Performance Applied to High-Tech Firms

Pay for performance can be shown to be appropriate for high-tech firms, if we agree that innovation and risk-taking can, indeed, be encouraged by highly variable rewards. In theory, at least, individual entrepreneurial behavior can be rewarded with individual incentives, but the design and delivery of the program is critical, and the failures recorded in the academic and practitioner literature are as numerous as the successes.

Merit pay systems have been shown to be correlated with employee satisfaction in a number of experiments and surveys (e.g., Heneman, Greenberger, & Strasser, 1988; Kanugo & Mendonce, 1988; Lawler, 1971; Locke, Feren, McCaleb, Shaw, & Denny, 1980; Nalbantian, 1987; Peck, 1984; Perry, Petrakis, & Miller, 1984; Yankelovich & Immerwahr, 1983). In some cases, the pay for performance study specifically correlated perception of equity to satisfaction (Dreher, 1981; Weiner, 1980). Jenkins and Lawler (1981) examined a situation in which the pay for performance design for hourly workers correlated to pay satisfaction in a program in which the workers participated in the pay program design process. In some programs that did not work well, one failure was that emeployees did not have the belief that the merit increase reflected true performance or vice versa (Hills, Scott, Markham, & Vest, 1987; Kahn & Sherer, 1990; Markham & Peterca, 1986; Russell et al., 1986). In a couple of cases, confidence in the performance rating was thought to decrease, over time (Allen & Rosenberg, 1986; Perry et al., 1989). Another problem was that the merit component of the total pay increase was likely too small (Perry & Petrakis, 1988). Finally, pay for performance may simply have not matched the culture of the organization (Birnbaum, 1983).

In order for a pay for performance system to have a chance of success, (1) the merit increase has to match the performance measurement system, (2) the employees must have a belief that the pay program is equitable, (3) the merit component of pay must be significant. As we had previously suggested, expectancy theory tells us that rewards, especially variable rewards, must be significant in order to have impact; merit pay increases are of little motivational value if the merit component of pay is very small. Lawler (1989) argued that a poorly designed pay for performance system is worse than not at all.

Learning theory offers the possibility that variable schedule rewards may provide an additional motivational tool. Nearly every business culture in the world has the tradition of the annual (12-month) wage increase. Certainly, very strong performance messages could be sent to individuals if the increase date were less than or greater than twelve months. Given expectation of a twelve-month review, a merit increase in less than twelve months can be a very clear positive message, and the delivery of this message is not particularly costly to the organization. However, messages sent through the pay system must be consistent; which is to say, an increase delivered in less than the expected time period should also be higher than an average increase amount or percentage. An increase that is delayed beyond twelve months, with the intent of sending a strong negative message, should then also be less than average; at this point, the employee should also be in counseling, to make certain the reasons for the action are known, and the employee understands how he or she can change behaviors in order to avoid future difficulties.

A Note on Performance Appraisal

Performance evaluation can serve to support the pay for performance ethic. At the same time it is one of the most problematical issues in the human resources practitioner's work life (Gehrman, 1984). The shortcomings and failure rates of performance appraisal systems are well documented (e.g., Lazer & Wikstrom, 1977). Lazer (1980) summarized five reasons why performance appraisal systems fail: (1) trying to do too many things with a single format, (2) lack of management support, (3) lack of job relatedness, (4) rater bias, and (5) difficulty in administration.

Behavior-based appraisal systems, including behaviorally anchored rating scales (Bernardin & Smith, 1981; Borman, 1975; Borman & Dunnette, 1975; Campbell et al., 1970; Campbell, Dunnette, Arvey, & Hekkervik, 1973; Smith & Kendall, 1963), behavioral observation scales (Latham, Fay, & Saari, 1979; Latham, Saari, & Fay, 1980; Latham & Wexley, 1977), and mixed standard scales (Arvey & Hoyle, 1974; Blanz & Ghiselli, 1972) seem to have shown the most promise in terms of reducing rating errors and in gaining employee and management confidence. The best performance appraisal systems, in fact, are built upon performance criteria that are (1) observable behaviors (Brumback, 1972; Kane & Lawler, 1979), (2) relevant to the job (Landy & Farr, 1983), and (3) significant. In addition to having the best chance of being accepted in the organization, such systems have the highest probability of being defensible in court (Cascio & Bernardin, 1981). In addition to reasonable criteria, successful programs seek to eliminate or reduce rater biases and rating errors. There is both good evidence and intuitive appeal for improving rater skills through formal training (Bernardin & Buckley, 1981; Borman, 1975; Landy & Farr, 1980, 1983;) and feedback (Davis & Mount, 1984).

Let us suggest that the broad system of performance evaluation consists of at least four subsystems, (1) formal performance appraisal, (2) the informal performance dialogue that presumably occurs between a good supervisor and his or her subordinates on a more-or-less daily basis, (3) the employee development program, and (4) feedback provided by the merit pay system. In this context, formal performance appraisal may be a relatively minor element, and certainly does not take the place of every-day open communication between supervisors and subordinates.

Pay for Contribution, for the Long Haul

The problems of potential turnover and the need to recruit specialized talent, both of which are particularly acute in high-tech firms, suggest the need for a sound, competitive base salary program, adequate benefits, and supportive human resources systems. Because of the seeming contradiction of high-tech environment requirements, between the need for an entrepreneurial spirit, on one hand, and the need for a stable pool of talents, on the other hand, a "pure" pay for performance system might not be quite possible. Some consideration should be given to pay for job holding and pay for membership, as well, to retain key individuals. As indicated above, "pay for contribution" combines pay for performance with recognition of the importance of the job to the organization (i.e., the opportunity the job affords the individual to deliver significant returns to the firm). This scheme may be even more appropriate, over the long term, for high-tech firms, than pay for performance, pay for job holding, or pay for membership; in fact, pay for contribution considers some elements of each. High-tech firms are likely to go in this direction as they become larger, more complex, and more mature. This does not mean however, that the performance component need be lost. The key to recognizing performance is still the willingness of management to apply a degree of variability to individual rewards. The pay for contribution ethic allows rewards for both the value of the job and the effort of the individual, within the scope of that job.

In a pay-for-contribution system, a high percentage of pay is tied to the job, and a much smaller percentage is tied to individual performance. It is entirely possible, however, that there can be a certain motivational leveraging of the performance component of pay, if it is true that pay has value in the recognition it provides (Krefting & Mahoney, 1977; Lawler, 1971). Therefore, there should be a kind of ceremonial flavor to awarding a merit increase, and the meaning of that increase, relative to the norm, should be fully exploited by management.

The Role of Job Analysis, Job Evaluation, and Job Pricing

Pay for contribution systems are generally built upon the use of job analysis and job evaluation to achieve internal equity (Belcher, 1974; Cascio, 1982; International Labour Organization, 1986; Madigan, 1985; Treiman, 1979); jobs are ranked or valued according to their relative contribution to the organization, so that a reasonable and equitable pay structure can be developed for those jobs.

Job evaluation is said to have the additional value of being able to deal with comparable worth issues (Arvey, Passino, & Lounsbury, 1977; Doverspike & Barrett, 1984; Doverspike, Carlisi, Barrett, & Alexander, 1983; Fraser, Cronshaw, & Alexander, 1984; Grams & Schwab, 1985; Mount & Ellis, 1987; Schwab & Wichern, 1983; Treiman, 1979; Treiman & Hartmann, 1981), at least, on the micro (single firm) level. In contrast, pure market pricing of jobs tends to perpetuate bias against female-dominated jobs (Brennan, 1984) and could lead to internal organizational dissatisfaction and a feeling of inequity. The management of high-tech or any firm should be working toward equity. It is probably impossible to achieve perfect equity in this world. However, the fact of knowing that equity is difficult to achieve does not absolve management from the responsibility of seeking equity.

Why Disclosed Pay Systems Are More Appropriate

The high-powered environment of high-tech firms not only demands "fairness" (equity) in compensation decisions by management, but also believe that things are being done fairly (Adams, 1963, 1965; Lawler, 1983). Compensation systems that are quite equitable in objective terms of market competitiveness, internal comparison of jobs, and rewarding individual performance, may still fail to motivate, if employees do not share, with management, that perception of equity; the failure may be not because of compensation design, but because of poor communications, or lack of trust. Secret compensation systems seem to be counterproductive, for this reason (Lawler, 1983). Furthermore, common sense suggests that if the firm truly has a quality compensation system, the system ought to be fully disclosed and adequately communicated, and management ought to take credit for doing a good job in compensation program design.

The Variable Pay Component of a Performance Rewards System

There is some evidence that bonus programs may be more powerful motivators than base salary merit pay systems (Finkelstein & Hambrick, 1988, 1989; Kahn & Sherer, 1990; Murphy, 1985; Schwab & Olson, 1988). In theory, bonus plans have the potential of meeting the tests of significance, in that funds can be accrued and concentrated in a single-payoff "event," and of being closely tied to performance, particularly if performance is explicitly defined as meeting certain measurable goals.

High-tech firms are characterized by products or processes that are leading edge and innovative. Innovation on the part of individuals is prized and may be measured. Equity in such firms is clearly not the same as "equality": greater differentiation of rewards may be more easily perceived as "equitable" than in traditional firms, where individual contribution is more difficult to identify. If group incentives are used, these may be more effective in small groups, such as project teams (Lawler, 1971). If individuals are operating in uncharted waters, agency theory might suggest that behavior-based reward systems are not appropriate (insofar as the correct behaviors are not yet defined), and that outcome-based systems might be better. In theory, then, more variable, more results-based rewards would be appropriate in the high-tech environment, and that rewards could be effectively delivered in a bonus format (Hardy, Orton, & Moffit, 1986).

A lengthy study at Du Pont (McNutt, 1990) developed some reasonable considerations in developing a variable pay system, which may be of interest to high-tech firms:

1. A compensation change should support, not drive a cultural change;
2. compensation should be tied to business success;
3. sharing in the success of the business can be an important tool for aligning employees' interest with those of the company, if employees perceive the compensation program as fair;
4. placing current compensation at risk can be a demotivator (on the other hand, placing future increases at risk can be a strong aligning and motivating tool; and
5. new pay systems must be continually tested for understanding, credibility, and fairness.

In truth, the Du Pont variable pay system, itself, has not yet been adequately tested. Some time may pass before management knows how well it has done in the design of their program. However, the conclusions reached in the research phase (interviews of employees and study of the pay programs of numerous other firms) are generally consistent with what theory might have led us to expect.

Employee Ownership Programs

Employee ownership is a relatively recent phenomenon, and it has stimulated some interest among researchers. Certain cases, like Weirton Steel, have caught the fancy of the business community and the general public, as well. An opportunity exists for greater organizational democracy, redistribution of organizational power, and participation in the economic future of the firm. However there seems to be often a gap between economic ownership and psychological "ownership;" which is to say, owner-workers see themselves more as "investors" than as active owners (French, 1987; Hammer & Stern, 1980). At least one study (Marsh & McAllister, 1981), however, reported the productivity of ESOP firms to be somewhat higher than other firms, and another (Rosen & Klein, 1983) found employee-owned firms grew more, in terms of number of employees. Cotton, Vollrath, Froggatt, Lengnick-Hall, and Jennings (1988) reviewed employee ownership as a special form of participation, and found a positive effect on both performance (defined in terms of absenteeism, turnover, work quality, and profit) and worker satisfaction. It may be that employee ownership, unto itself, does not guarantee improved performance, but that when employee ownership is combined with a culture of participation and worker autonomy it can be a powerful incentive to perform (e.g., Kirkpatrick, 1988).

Are Participative Styles Worth Considering?

Participative managment was a primary characteristic of the so-called "human relations movement" that grew out of the Hawthorne studies (Roethlisberger & Dixon, 1939) and the writings of George Elton Mayo (1933, 1945). Later, participative styles were also a key feature of the "human resources movement," led by Argyris (1957), Herzberg et al. (1959), McGregor (1960), Likert (1961), Vroom (1964), Miles (1965), and others. A number of empirical studies and literature reviews have since claimed success for participation, autonomy, and reduced supervision, as opposed to tight bureaucratic controls (e.g., Coch & French, 1948; Dessler, 1976; Erez & Arad, 1986; Jackson, 1983; Lawler, 1983; Locke & Schweiger, 1979; Miller & Monge, 1986; Nicholson & Goh, 1983; Nightingale, 1987). Cotton et al. (1988) reviewed the literature, using a multidimensional approach, and found positive effects for (1) participation in work decisions and performance, (2) informal participation and performance, (3) informal participation and satisfaction, (4) employee ownership and performance, and (5) employee ownership and satisfaction.

The reason for introducing this material into a discussion of compensation is to suggest the strong intuitive appeal that, if one is arguing for employee motivation through variable pay, might not the effects of variable pay be enhanced by increasing employee control over the results that determine variable pay outcomes? It is effective to, for example, pay a bonus on the performance of the organization, if employees feel they have little influence on performance results, but have the feeling, instead, that good and bad financial outcomes are mostly a function of top management decisions? We would suggest that the compensation system ought to fit the culture of the organization, and that compensation funds spent in a fashion that does not reflect the management style, might well be money wasted. In a top-down, tight-control climate, do pay incentives make sense, other than with respect to those who are at the very top, and who do have the real control?

Integrating Pay for Performance Into a Broader HRM Strategy

Lawler's congruity theory (1976, 1983, 1989) suggested that pay systems must be in concert with organizational factors, management style, the climate of the organization, group size, and so forth. This seems to be particularly important to high-tech firms, whose characteristics may be quite different from other organizations. A pay system where individual incentives were emphasized might not be consistent with small group decision processes. Neither individual incentives nor small group incentives would make sense if the relative autonomy of the individual or small group was low. In the case of an organization where an individual's contribution could clearly be defined in terms of measurable outcomes, pay for performance could be implemented in the form of a bonus plan, with the payout based on achievement of planned objectives. Where outcomes are not easily measures, a base salary merit raise system, with performance defined in terms of a behavior-based performance appraisal program, might be more appropriate. In most situations, both circumstances may be true, in which case, the combination of a bonus plan for measureable accomplishments plus a base salary merit increase to reinforce desired but more subjectively-judged behaviors, might be the best design. If turnover is atypical of high-tech firms (Balkin & Gomez-Mejia, 1984), it may not be possible to use a highly at-risk contingency pay program. Base salaries would have to be fairly competitive by themselves, and bonus awards would be clearly above and beyond an attractive base salary.

Although high-tech firms are generally defined as being in a growth mode, there may be considerable variability from firm to firm as to temporal position in the product life cycle, size of the firm, and sophistication of internal control processes. As firms mature, become more stable, grow in size, and establish integrative management systems, agency theory would suggest shifting from an outcome-based reward system to a behavior-based system. Reward systems are more likely to be behavior-based if the principal's information about the internal workings of the organization are good, if desirable behaviors can be identified and defined, and when longer-term growth goals replace short-term goals (Eisenhardt, 1989).

CONCLUSION

We have argued the importance of understanding motivation theory in order to correctly design compensation systems. In the compensation business, we don't have the luxury of conclusive empirical evidence, nor are we all in agreement about the meaning of experiential findings. The confounding variables are, alas, numerous and powerful. We also may not have the luxury of experimenting, and, even if we do, it may take some number of pay plan cycles (i.e., years) before we know whether or not a pay program is working well. Therefore, we have to look to theory to find a firm foundation upon which to build a compensation system.

We have thought about pay for performance, as being quite appropriate to the entrepreneurial culture of high-tech firms. Over the longer term, however, there may be a steadying effect to be gained from pay for contribution, while not entirely losing the high performance ethic. Finally, we believe any pay system must reinforce and be consistent with the business strategy of the firm, organizational culture and climate, and the other human resources management programs in place.

We hope this piece does stimulate thought about compensation system design in high-tech firms. Each firm's management must know what they are all about, what they stand for, and what they believe, and make decisions accordingly. We, finally, might hope to stimulate thought about compensation research, so that some of the gaps in empirical knowledge might some day be closed.

REFERENCES

Adams, J.S. (1963). Toward an understanding of inequity. *Journal of Abnormal and Social Psychology, 67,* 422-436.

Adams, J.S. (1965). Inequity in social exchange. In L. Berkowitz (Ed.), *Advances in experimental social psychology* (Vol. 2, pp. 267-299). New York: Academic Press.

Allen, P., & Rosenberg, S. (1986). An assessment of merit pay administration under New York City's managerial performance evaluation system: Three years of experience. *Public Personnel Management, 15*(3), 297-309.

Andrews, I.R. (1967). Wage inequity and job performance: An experimental study. *Journal of Applied Psychology, 51,* 39-45.

Argyris, C. (1957). *Personality and organization.* New York: Harper & Brothers.

Arvey, R.D., & Hoyle, J.C. (1974). A Guttman approach to the development of behaviorally based rating scales for systems analysts and programmer/analysts. *Journal of Applied Psychology, 59,* 61-68.

Arvey, R.D., Passino, E.M., & Lounsbury, J.W. (1977). Job analysis results as influenced by sex of incumbent and sex of analyst. *Journal of Applied Psychology, 62,* 411-416.

Balkin, D.B., & Gomez-Mejia, L.R. (1984). Determinants of R-and-D compensation strategies in the high tech industry. *Personnel Psychology, 37,* 635-650.

Barnard, C.I. (1938). *The functions of the executive.* Cambridge, MA: Harvard University Press.

Bates, Marsh W. (1981). *Motivating performance through cash compensation.* 1981 Hay Compensation Conference Proceedings (pp. 7-14). Philadelphia, PA: Hay Associates.

Belcher, D.W. (1974). *Compensation administration.* Englewood Cliffs, NJ: Prentice-Hall.

Berger, C.J., Cummings, L.L., & Heneman, H.G. (1975). Expectancy theory and operant conditioning predictions of performance under variable ratio and continuous schedules of reinforcement. *Organizational Behavior and Human Performance, 14,* 227-243.

Bernardin, H.J., & Buckley, M.R. (1981). Strategies in rater training. *Academy of Management Review, 6,* 205-212.

Bernardin, H.J., & Smith, P.C. (1981). A clarification of some issues regarding the development and use of behaviorally anchored rating scales (BARS). *Journal of Applied Psychology, 66,* 458-463.

Birnbaum, M.H. (1983). Perceived equity of salary policies. *Journal of Applied Psychology, 68*(1), 49-59.

Blanz, F., & Ghiselli, E.E. (1972). The mixed standard scale: A new rating system. *Personnel Psychology, 25,* 185-199.

Borman, W.C. (1975). Effects of instructions to avoid halo error on reliability and validity of performance evaluation ratings. *Journal of Applied Psychology, 60,* 556-560.

Borman, W.C., & Dunnette, M.D. (1975). Behavior-based versus trait-oriented performance ratings: An empirical study. *Journal of Applied Psychology, 60,* 561-565.

Brennan, E.J. (1984, October). Sex discrimination and comparable worth. *Personnel Journal,* 56-59.

Brumback, G.A. (1972). A reply to Kavenaugh. *Personnel Psychology, 25,* 567-572.

Campbell, J.P., Dunnette, M.D., Arvey, R.D., & Hekkervik, L.V. (1973). The development and evaluation of behaviorally based rating scales. *Journal of Applied Psychology, 57,* 15-22.

Campbell, J.P., Dunnette, M.D., Lawler, E.E., III, & Weick, K.E. (1970). *Managerial behavior, performance and effectiveness.* New York: McGraw-Hill.

Cascio, W.F. (1982). *Applied psychology in personnel management.* Reston, VA: Reston Publishing.

Cascio, W.F., & Bernardin, H.J. (1981). Implications of performance appraisal litigation for personnel decisions. *Personnel Psychology, 34,* 211-226.

Coch, L., & French, J.R.P., Jr. (1948). Overcoming resistance to change. *Human Relations,* 512-532.

Conlon, E., & Parks, J. (1988). The effects of monitoring and tradition on compensation arrangements: An experiment on principal/agent dyads. In F. Hoy (Ed.), *Best papers proceedings.* Anaheim, CA: Academy of Management.

Cook, K.S. (1975). Expectations, evaluations and equity. *American Sociological Review, 40,* 372-388.

Cotton, J.L., Vollrath, D.A., Froggatt, K.L., Lengnick-Hall, M. L., & Jennings, K.R. (1988). Employee participation: Diverse forms and different outcomes. *Academy of Management Review, 13*(1), 8-22.

Dachler, H.P., & Hulin, C.L. (1969). A reconsideration of the relationship between satisfaction and judged importance of environmental and job characteristics. *Organizational Behavior and Human Performance, 4,* 252-262.

Daniel, T.L., & Esser, J.K. (1980). Intrinsic motivation as influenced by rewards, task interest, and task structure. *Journal of Applied Psychology, 65*(5), 566-573.

Davis, B.L., & Mount, M.K. (1984, March). Design and use of a performance appraisal feedback system. *Personnel Administrator,* 91-97.

Dessler, G. (1976). *Organization and management: A contingency approach.* Englewood Cliffs, NJ: Prentice-Hall.

Doverspike, D., & Barrett, G.V. (1984). An internal bias analysis of a job evaluation instrument. *Journal of Applied Psychology,69*(4), 648-662.

Doverspike, D., Carlisi, A.M., Barrett, G.V., & Alexander, R.A. (1983). Generalizability analysis of a point-method job evaluation instrument. *Journal of Applied Psychology, 68*(3), 476-483.

Downey, H.K., Chacko, T.I., & McElroy, J.C. (1979). Attribution of the "causes" of performance: A constructive, quasi- longitudinal replication of the Staw (1975) study. *Organizational Behavior and Human Performance, 24,* 287-299.

Dreher, G.F. (1981). Predicting the salary satisfaction of exempt employees. *Personnel Psychology, 34,* 579-589.

Drucker, P. (1954). *The practice of management.* New York: Pitman.

Dyer, L., Schwab, D.P., & Theriault, R.L. (1976). Managerial perceptions regarding salary increase criteria. *Personnel Psychology, 29,* 233-242.

Dyer, L., & Theriault, R.L. (1976). The determinants of pay satisfaction. *Journal of Applied Psychology, 61,* 596-604.

Ehrenberg, R.G. (1990, February). Introduction: Do compensation policies matter? *Industrial and Labor Relations Review,* 43, 3S-12S.

Ehrenberg, R.G., & Milkovich, G.T. (1987). Compensation and firm performance. In M. Kleiner et al. (Eds.), *Human resources and the performance of the firm* (pp. 87-132). Madison, WI: Industrial Relations Research Association.

Eisenhardt, K.M. (1985). Control: Organizational and economic approaches. *Management Science, 31,* 134-149.

Eisenhardt, K.M. (1988). Agency- and institutional-theory explanations: The case of retail sales compensation. *Academy of Management Journal, 31*(3), 488-511.

Eisenhardt, K.M. (1989). Agency theory: An assessment and review. *Academy of Management Review, 14*(1), 57-74.

Erez, M., & Arad, R. (1986). Participative goal-setting: Social, motivational, and cognitive factors. *Journal of Applied Psychology, 71*(4), 591-597.

Etzioni, A. (1961). *A comparative analysis of complex organizations on power, involvement, and their correlates.* New York: The Free Press.

Evans, W.A. (1970). Pay for performance: Fact or fable. *Personnel Journal, 49*(9), 726-731.

Festinger, L. (1957). *A theory of cognitive dissonance.* Evanston, IL: Row, Peterson.

Finkelstein, S., & Hambrick, D.C. (1988). Chief executive compensation: A synthesis and reconciliation. *Strategic Management Journal, 9,* 543-558.

Finkelstein, S., & Hambrick, D.C. (1989). Chief executive compensation: The intersection of markets and political processes. *Strategic Management Journal, 10,* 121-134.

Finn, R.H., & Lee, S.M. (1972). Salary equity: It's determination, analysis and correlates. *Journal of Applied Psychology, 56,* 292-293.

Fossum, J.A., & Fitch, M.K. (1985). The effects of individual and contextual attributes on the sizes of recommended salary increases. *Personnel Psychology, 36,* 587-602.

Foster, K., & Lynn, K.V. (1978). Dividing up the "merit" increase pie for management. *Personnel Administrator, 23*(5), 379-409.

Fraser, S.L., Cronshaw, S.F., & Alexander, R.A. (1984). Generalizability analysis of a point-method job evaluation instrument: A field study. *Journal of Applied Psychology, 69*(4), 643-647.

French, J. L. (1987). Employee perspectives on stock ownership: Financial investment or mechanism for control? *Academy of Management Review, 12*(3), 427-435.

Garland, H. (1973). The effects of piece-rate underpayment and overpayment on job performance. *Journal of Applied Psychology, 57,* 325-334.

Gehrman, D.B. (1984). Beyond today's compensation and performance appraisal systems. *Personnel Administrator, 29*(3), 21-33.

Gomez-Mejia, L.R., & Welbourne, T.M. (1988). Compensation strategy: An overview and future steps. *Human Resources Planning, 11,* 173-189.

Gould, S. (1979). An equity-exchange model of organizational involvement. *Academy of Management Review, 41*(1), 53-62.

Grams, R., & Schwab, D.P. (1985). An investigation of systematic gender-related error in job evaluation. *Academy of Management Journal, 28*(2), 279-290.

Gupta, N., Jenkins, G.D., Jr., Curington, W.P., Clements, C., Doty, D.H., Schweizer, T., & Teutsch, C.H. (1986). *Exploratory investigations of pay-for-knowledge systems.* Washington, DC: Bureau of Labor-Management Relations and Cooperative Programs.

Hall, D.T., & Nougaim, K.E. (1968). An examination of Maslow's need hierarchy in an organizational setting. *Organizational Behavior and Human Performance, 3,* 12-22.

Hammer, T.H., & Stern, R.N. (1980). Employee ownership: Implications for the organizational distribution of power. *Academy of Management Journal, 23*(1), 78-100.

Hardy, J.W., Orton, B.B., & Moffit, J.W. (1986, November). Bonus systems do motivate. *Management Accounting,* 58-61.

Heneman, H.G., III., & Schwab, D.P. (1972). Evaluation of research on expectancy theory predictors of employee performance. *Psychological Bulletin, 78,* 1-9.

Heneman, R.L., Greenberger, D.B., & Strasser, S. (1988). The relationship between pay-for-performance perceptions and pay satisfaction. *Personnel Psychology, 41,* 745-759.

Herzberg, F., Mausner, B., & Snyderman, B.B. (1959). *The motivation to work.* New York: Wiley.

Hills, F.S. (1979). The pay for performance dilemma. *Personnel, 56*(5), 23-31.

Hills, F.S., Scott, K.D., Markham, S.E., & Vest, M.J. (1987, September). Merit pay: Just or unjust desserts? *Personnel Administrator,* 53-59.

Homans, G.C. (1961). *Social behavior: Its elementary forms.* New York: Harcourt, Brace and World.

Hull, C.L. (1943). *Principles of behavior.* New Haven, CT: Yale University Press.

Hull, C.L. (1951). *Essentials of behavior.* New Haven, CT: Yale University Press.

Hull, C.L. (1952). *A behavior system.* New Haven, CT: Yale University Press.

Ilgen, D.R., Nebeker, D.M., & Pritchard, R.D. (1981). Expectancy theory measures: An empirical comparison in an experimental simulation. *Organizational Behavior and Human Performance, 28,* 189-223.

International Labour Organization. (1986). *Job evaluation.* Geneva: International Labour Office.

Jackson, S. E. (1983). Participation in decision making as a strategy for reducing job-related strain. *Journal of Applied Psychology, 68*(1), 3-19.

Jenkins, G.D., & Lawler, E.E., III. (1981). Impact of employee participation in pay plan development. *Organizational Behavior and Human Performance, 28,* 111-128.

Kahn, L.M., & Sherer, P.D. (1990, February). Contingent pay and managerial performance. *Industrial and Labor Relations, 43,* 107S-120S.

Kane, J.S., & Lawler, Edward E., III. (1979). Performance appraisal effectiveness: Its assessment and determinants. In B.M. Staw (Ed.), *Research in organizational behavior* (Vol. 1., pp. 476-483). Greenwich, CT: JAI Press.

Kanungo, R.N., & Mendonca, M. (1988, Fall). Evaluating employee compensation. *California Management Review,* 23-39.

Kearney, W.J. (1985). Pay for performance? Not always. In M. Beer & B. Spector (Eds.), *Readings in HRM* (pp. 425-433). New York: The Free Press.

Kessler, J.J., & Winer, J. (1972). Self consistency and inequity dissonance as factors in undercompensation. *Organizational Behavior and Human Performance, 8,* 456-466.

Kirkpatrick, D. (1988, December 5). How the workers run Avis better. *Fortune,* pp. 103-114.

Kopelman, R.E., & Reinharth, L. (1982). Research results: The effect of merit-pay practices on white collar performance. *Compensation Review, 14*(4), 30-40.

Krefting, L.A., & Mahoney, T.A. (1977). Determining the size of a meaningful pay increase. *Industrial Relations, 16*(1), 83-93.

Landy, F.J., & Farr, J.L. (1980). Performance rating. *Psychological Bulletin, 87,* 72-107.

Landy, F.J., & Farr, J.L. (1983). *The measurement of work performance: Methods, theory and applications.* New York: Academic Press.

Langer, W.C. (1937). *Psychology and human living.* New York: Appleton-Century Crofts.

Latham, G.P., Fay, C., & Saari, L. (1979). The development of behavioral observation scales for appraising the performance of foremen. *Personnel Psychology, 32,* 299-311.

Latham, G.P., Saari, L., & Fay, C. (1980). BOS, BES, and baloney: Raising Kane with Bernardin. *Personnel Psychology, 33,* 815-821.

Latham, Gary P., & Wexley, K.N. (1977). Behavioral observation scales for performance appraisal purposes. *Personnel Psychology, 30,* 255-268.

Lawler, E.E., III. (1971). *Pay and organizational effectiveness: A psychological view.* New York: McGraw-Hill.

Lawler, E.E., III. (1972). Secrecy and the need to know. In M. Dunnette, R. House, & H. Tosi (Eds.), *Readings in managerial motivation and compensation* (pp. 362-371). East Lansing, MI: Michigan State University Press.

Lawler, E.E., III. (1976). Control systems in organizations. In M.D. Dunnette (Ed.), *Handbook of industrial and organizational psychology* (pp. 1247-1292). Chicago: Rand-McNally.

Lawler, E.E., III. (1983). *Pay and organizational development.* New York: Addison-Wesley.

Lawler, E.E., III. (1989). Pay for performance: A strategic analysis. In L.R. Gomez-Mejia (Ed.), *Compensation and benefits* (pp. 136-181). Washington, DC: The Bureau of National Affairs.

Lawler, E.E., III, & Ledford, G.E., Jr. (1985, September). Skill-based pay: A concept that's catching on. *Personnel,* pp. 30-37.

Lawler, E.E., III, & Suttle, J.L. (1972). A causal correlation test of the need hierarchy concept. *Organizational Behavior and Human Performance, 2,* 265-275.

Lawler, E.E., III, & Suttle, J.L. (1973). Expectancy theory and job behavior. *Organizational Behavior and Human Performance, 9,* 482-503.

Lazer, R.I. (1980, July). Performance appraisal: What does the future hold? *Personnel Administrator,* pp. 69-73.

Lazer, R.I., & Wikstrom, W.S. (1977). *Appraising managerial performance: Current practices and future directions.* New York: The Conference Board.

Lewin, K. (1947). Group decision and social change. In T.M. Newcomb & E.L. Harley (Eds.), *Readings in social psychology* (pp. 330-344). New York: Henry Holt.

Likert, R. (1961). *New patterns of management.* New York: McGraw-Hill.

Locke, E.A., Feren, D.B., McCaleb, V.M., Shaw, K.N., & Denny, A. (1980). The relative effectiveness of four methods of motivating employee performance. In K.D. Duncan, M.M. Gruneberg, & D. Wallas (Eds.), *Changes in work life* (pp. 363-388). Chichester, UK: Wiley.

Locke, E.A., & Schweiger, D.M. (1979). Participation in decision-making: One more look. *Research in Organizational Behavior, 1,* 265-339.

Luthans, F., & Fox, M.L. (1989, March). Update on skill-based pay. *Personnel,* pp. 26-31.

Madigan, R.M. (1985). Comparable worth judgments: A measurement properties analysis. *Journal of Applied Psychology, 70*(1), 137-147.

March, J.G., & Simon, H.A. (1958). *Organizations.* New York: Wiley.

Markham, S.E., & Peterca, R. (1986). *The pay-for-performance dilemma revisited: An empirical example of the importance of group effects.* Paper presented at the annual meeting of the Academy of Management, Chicago.

Marsh, T., & McAllister, D. (1981). ESOP tables: A survey of companies with employee stock ownership plans. *Journal of Corporation Law, 6,* 551-624.

Maslow, A.H. (1943). A theory of human motivation. *Psychological Review, 50,* 370-396.

Maslow, A.H. (1954). *Motivation and personality.* New York: Harper.

Mawhinney, T.C. (1979). Intrinsic x extrinsic work motivation: Perspectives from behaviorism. *Organizational Behavior and Human Performance, 24,* 411-440.

Mayo, G.E. (1933). *The human problems of an industrial civilization.* New York: MacMillan.

Mayo, G.E. (1945). *The social problems of an industrial civilization.* Cambridge, MA: Harvard University Press.

McGregor, D.M. (1960). *The human side of enterprise.* New York: McGraw-Hill.

McNutt, R.P. (1990). Achievement pays off at Du Pont. *Personnel, 67*(6), 5-10.

Meyer, H.H. (1975). The pay for performance dilemma. *Organizational Dynamics, 3*(3), 39-50.

Miles, R.E. (1965, July-August). Human relations or human resources? *Harvard Business Review*, 149-161.

Milkovich, G.T. (1988). A strategic perspective in compensation management. *Research in personnel and human resources management, 6*, 263-288.

Miller, K.I., & Monge, P.R. (1986). Participation, satisfaction, and productivity: A meta-analytic review. *Academy of Management Journal, 29*(4), 727-753.

Mount, M.K., & Ellis, R.A. (1987). Investigation of bias in job evaluation ratings of comparable worth study participants. *Personnel Psychology, 40*(1), 85-96.

Murphy, K. (1985). Corporate performance and managerial remuneration. *Journal of Accounting and Economics, 7*, 11-42.

Nash, J.F., Jr. (1950). The bargaining problem. *Econometrica, XXI*, 128-140.

Nalbantian, H. (1987). *Incentives, cooperation, and risk sharing*. Totowa, NJ: Rowman and Littlefield.

Nicholson, P.J., Jr., & Goh, S.C. (1983). The relationship of organization structure and interpersonal attitudes to role conflict and ambiguity in different work environments. *Academy of Management Journal, 26*(1), 148-155.

Nightingale, D.V. (1981). Participation in decision-making: An examination of style and structure and their effects on member outcomes. *Human Relations, 34*, 1-12.

Odiorne, G. (1964). *Management by objectives*. New York: Harper.

O'Reilly, C.A., III, & Caldwell, D.F. (1980). Job choice: The impact of intrinsic and extrinsic factors on subsequent satisfaction and commitment. *Journal of Applied Psychology, 65*(5), 559-565.

Oviatt, B.M. (1988). Agency and transaction cost perspectives on the manager-shareholder relationship: Incentives for congruent interests. *Academy of Management Review, 13*(2), 214-225.

Patchen, M. (1961). *The choice of wage comparisons*. Englewood Cliffs, NJ: Prentice-Hall.

Peck, C. (1984). *Pay for performance: The interaction of compensation & performance appraisal*. New York: The Conference Board.

Pearce, J.L., Stevenson, W.B., & Perry, J.L. (1985). Managerial compensation based on organizational performance: A time series analysis of the effects of merit pay. *Academy of Management Journal, 28*(2), 261-278.

Perry, J.L., & Petrakis, B.A. (1988, January/February). Can pay for performance succeed in government? *Public Personnel Management, 17*(4), 359-367.

Perry, J.L., Petrakis, B.A., & Miller, T.K. (1989, January/February). Federal merit pay, round II: An analysis of the performance management and recognition system. *Public Administration Review*, 29-37.

Porter, L.W., & Lawler, E.E., III. (1965). Properties of organization structure in relation to job behavior. *Psychological Bulletin, 64*, 23-57.

Pfeffer, J., & Davis-Blake, A. (1987). Understanding organizational wage structure: A resource dependence approach. *Academy of Management Journal, 30*(3), 437-455.

Roethlisberger, F.J., & Dixon, W.J. (1939). *Management and the worker*. Cambridge, MA: Harvard University Press.

Rosen C., & Klein, K. (1983). Job-creating performance of employee-owned firms. *Monthly Labor Review, 106*(8), 15-19.

Russell, C.J., Gallagher, D.G., & Tokesky, G.C. (1986). *An examination of merit pay at managerial and professional levels in a pre-raise post-raise design*. Paper presented at the annual meeting of the Academy of Management, Chicago.

Saari, L.M., & Latham, G.P. (1982). Employee reactions to continuous and variable ratio reinforcement schedules involving a monetary incentive. *Journal of Applied Psychology, 67*(4), 506-508.

Savery, L.K. (1988). Comparison of managerial and non-managerial employees' desired and perceived motivators and job satisfaction levels. *Leadership and Organizational Development, 9*(1), 17-22.

Schackle, G.L.S. (1949). *Expectation in economics.* Cambridge: Cambridge University Press.

Schneier, C.E. (1974). Behavior modification in management: A Review and critique. *Academy of Management Journal, 3,* 528-548.

Schuster, J.R. (1985, May). Compensation plan design. *Management Review, 21-25.*

Schwab, D.P., & Dyer, L. (1973). The motivational impact of a compensation system on employee performance. *Organizational Behavior and Human Performance, 9,* 215-225.

Schwab, D.P., & Grams, R. (1985). Sex-related errors in job evaluation: A "real world" test. *Journal of Applied Psychology, 70*(3), 533-539.

Schwab, D.P., & Olsen, C.A. (1988). Pay-performance relationships as a function of pay for performance policies and practices. *Academy of Management Best Papers Proceedings,* 287-291.

Schwab, D.P., & Wichern, D.W. (1983). Systematic bias in job evaluation and market wages: Implications for the comparable worth debate. *Journal of Applied Psychology, 68*(1), 60-69.

Skinner, B.F. (1953). *Science and behavior.* New York: MacMillan.

Smith, P.C., & Kendall, L.M. (1963). Retranslation of expectations: An approach to the construction of unambiguous anchors for rating scales. *Journal of Applied Psychology, 47,* 149-155.

Staw, B.M. (1975). Attribution of the "causes" of performance: A new alternative interpretation of cross-sectional research on organizations. *Organizational Behavior and Human Performance, 13,* 414-432.

Treiman, D.J. (1979). *Job evaluation: An analytic review.* Washington, DC: National Academy of Sciences.

Treiman, D.J., & Hartman, H.I. (1981). *Women, work, and wages: Equal pay for jobs of equal value.* Washington, DC: National Academy Press.

Ungson, G.R., & Steers, R.M. (1984). Motivation in executive compensation. *Academy of Management Review, 9*(2), 313-323.

Vroom, V.H. (1964). *Work and motivation.* New York: Wiley.

Wahba, M.A., & Birdwell, L.G. (1976). Maslow reconsidered: A review of research on the need hierarchy theory. *Organizational Behavior and Human Performance, 15,* 212-239.

Weiner, N. (1980). Determinants and behavioral consequences of pay satisfaction: A comparison of two models. *Personnel Psychology, 33,* 741-757.

Wernimont, P.F., & Fitzpatrick, S. (1972). The meaning of money. *Journal of Applied Psychology, 56,* 226.

Williams, J. C. (1982). *Human behavior in organizations.* Cincinnati, OH: Southwestern.

Yankelovich, D., & Immerwahr, J. (1983, September 5). Let's put the work ethic to work. *Industry Week,* pp. 33-35.

Young, O.R. (1975). The bargainer's calculus. In O.R. Young (Ed.), *Bargaining: Formal theories of negotiation.* Urbana, IL: University of Illinois Press.

Yukl, G.A., Wexley, K.N., & Seymore, J.D. (1972). Effectiveness of pay incentives under variable ratio and continuous reinforcement schedules. *Journal of Applied Psychology, 56,* 19-23.

Zadeck, S., & Smith, P.C. (1968). A psychophysical determination of equitable payment: A methodological study. *Journal of Applied Psychology, 52,* 343-347.

PART II

STAFFING

RECRUITING AND HIRING THE HIGH-TECHNOLOGY PROFESSIONAL:

TRENDS AND FUTURE DIRECTIONS

Garth Coombs, Jr. and Joseph G. Rosse

It has been repeated often that high-technology firms live and die by the quality of their technical personnel. The types of innovative activities which are the lifeblood of the high-tech venture rely on availability of scientists and engineers with the technical proficiency to work at the "cutting edge" of their technologies and to push that edge ever outward in search of the next breakthrough. Supply of these technical personnel in many specialty areas falls far short of the level of demand from both high-tech and traditional firms in the labor market. When combined with the pace of change, need for quick reaction time, and the high profile that high-technology industry has come to have, a somewhat unique human resources (HR) arena has developed that is attracting increasing interest among researchers into human resource practices.

The goal of this chapter is to examine the current state of research into the recruitment and hiring processes utilized by high-technology (HT) firms with respect to the technical professionals they require. This restricts our focus in

Human Resource Strategy in High Technology.
Advances in Global High-Technology Management, Volume 1, pages 91-107.

ISBN: 1-55938-489-1

several ways. First, we examine research related to a narrow segment of the human resource management activities possible for consideration. Other papers in this volume have as their foci such issues as compensation, turnover, and training and development. We will attempt to minimize overlaps with such issues. Second, we look at the technical professionals—the scientists and engineers—and to some extent the managers of such employees. This focus excludes such groups as production workers and nontechnical employees in the HT organization. This focus is taken because there is general agreement that the technical personnel are perceived as critical to survival in the HT arena. Moreover, these employees are viewed as fundamentally different from other types of employees, an issue that will be addressed later in this paper. After discussing the state of current research, we will propose areas which may prove fruitful for future research efforts.

BACKGROUND

Anderson and Kleingartner (1987) comment, in the introductory chapter to their book, on the widely held belief that HT firms utilize innovative human resource practices and the large ripple effect this might have on more traditional firms. They go on, however, to challenge many of the conventional wisdoms and point out that:

> there are few reliable estimates of the prevalence or effects of innovative compensation and other human resource management practices in the industry. (p. 6)

Kleingartner and Mason (1986) attribute much of the focus on innovative human resource practices in HT to popular press attention on start-up firms.

While researchers have focused on compensation issues in HT (cf. Balkin & Gomez-Mejia, 1985; Gomez-Mejia & Welbourne, 1990; Milkovich, 1987), less attention has been paid to other human resource management practices. However, given the criticality of technical personnel to the success of the HT firm and the much commented upon shortfall in supply of such individuals, it would seem an important task to better understand the full range of HR practices used to attract and select the best possible technical personnel. It has been stated that traditional personnel management techniques fail in HT (Bleicher, Bleicher, & Paul, 1983), but there is a dearth of empirical evidence to support this contention or to indicate what succeeds.

Anderson and Kleingartner (1987) suggest that human resource practices in HT stem from the needs of management to (1) recruit technical professionals in tight labor markets, (2) develop and maintain their commitment to the firm, (3) provide incentives and job security, and (4) encourage productivity in innovation and the development of new products. What occurs in HT has important implications because of the high profile these industries have and the impacts they will have as trend setters for non-HT industries (Bestuzhev-Lada, 1987; Hayes & Abernathy, 1982). According to Belous (1987), this is true despite the fact that the HT sector does not employ a large percentage of the workforce. Belous also proposes that the central issue in the tightness of the HT labor market may be one of the quality of technical personnel available rather than the overall quantity available. This is debatable in that, for at least some key technical specializations, demand far outstrips the supply of even minimally qualified personnel (Griggs & Manring, 1985; Grissom & Lombardo, 1985; Rozen, 1985). Recent years have seen shakeouts in HT firms as some have failed, others have downsized, and still others have merged or have been acquired (Anderson & Kleingartner, 1987). This means that at times groups of personnel have gone into the labor market en masse, creating at least localized oversupplies of some types of HT workers. For the most part, however, it has been the production workers who have been most adversely affected, for the professionals have many "suitors" waiting in the wings either locally or nationally.

HIGH TECHNOLOGY AND PROFESSIONALISM

High-technology workers, and especially the technical specialists, are perceived to be a new breed (Anderson & Kleingartner, 1987). They appear to have different characteristics, expectations, and career paths than their peers in more traditional industries. The central issue which sets them apart is the concept of professionalism. Anderson and Kleingartner define "professionalism" in terms of the skill level of the worker, the type of work, and the ways in which the work is done. While Anderson and Kleingartner refer to scientists, engineers, and techicians collectively as professional workers, Kerr and Von Glinow (1977) assert that scientists and engineers, the two primary groupings of technical personnel, differ significantly on key attributes and should not be lumped together in research studies. Badawy (1971, 1988) agrees and indicates that engineers in traditional settings tend to be more career and organizationally oriented than the more professionally oriented scientists.

High-technology settings form a different microcosm, however, one in which engineers are treated like professionals (Hybels & Barley, 1988). It has been asserted by Hybels and Barley (1988) that progressive human resource policies that developed initially as a coping response to volatile and competitive labor markets for technical personnel, have contributed to the increasing professionalization of those personnel. In essence, HT employers are creating a self-fulfilling prophecy for technical professionals. As has been the case with innovative compensation practices, one firm's innovative HR practice diffuses rapidly and becomes applied industry wide. Later applicants entering the market expect similar treatment and come to consider such practices the minimal standard they should expect. Anderson and Kleingartner (1987) feel that educational socialization also contributes to the development of professional expectations, laying part of the responsibility at the doorstep of the universities that produce the scientists and engineers that HT demands.

The consequences of professionalization need attention. First, it is assumed that professionals maintain their loyalty more to the profession than to the employing firm (Turbin & Rosse, 1990). This reduced commitment to the firm is likely a contributory factor in the notoriously high turnover rates associated with HT enclaves and the relative ease of raiding employees from a competitor. Second, the professionalized employees may come to look down on technical personnel in other areas of the firm, hampering the level of integration needed to effectively develop and capitalize on innovative ideas. Third, there may be conflicts between organizational objectives and professional goals and values, leading to reduced individual and organizational performance.

Gerpott and Domsch (1985) take a group level view of the issue of professionalism. They posit that views on professionalism and managerialism in the United States may erode U.S. competitive advantages in technology management, and they hold the human resource practices responsible for many of the problems with technology management in the United States. By comparison to German firms, U.S. R&D oriented firms often split the technical and managerial elements by bringing in "professional" managers without technical expertise instead of promoting technical personnel into managerial positions. Gerpott and Domsch cite studies that indicate that technical professionals perform better when managed by technically expert R&D managers. The U.S. tendency to favor generalist managers may inhibit development of cohesion in the R&D hierarchy and to the maintenance of technical expertise.

In sharp contrast to this view, Sherman (1989) points out that the technical personnel's decision to quit is often related to managerial behavior. One might infer that those filling managerial positions without sufficient development of management skills become a source of dissatisfaction for their subordinates. Turbin and Rosse (1990) similarly assert that a key obstacle in attracting and retaining technical personnel is the poor managerial skill of scientists and engineers promoted into management slots. Moreover, many of those promoted are dissatisfied with being moved away from their technical roots and thus may not respond well to attempts to develop their managerial skills. Badawy (1988) also notes the dangers of promoting technically excellent engineers and scientists into managerial roles to which they are ill-suited. He comments that the firm in many cases both loses a first-class engineer and gains a second-rate technical manager who may do more harm than good. These observations seem to contradict the assertions of managerialism made by Gerpott and Domsch, however, they note that nondiversified firms tend to take a more specialist approach than do diversified firms. High technology firms, especially younger firms, fall more within the former category and therefore may not be as susceptible to the criticisms Gerpott and Domsch level as do more traditional R&D organizations.

While characteristics of management careers in HT have been described (Kanter, 1984), little research has focused on the recruitment and selection of these technical managers. In their survey of HT human resource managers, Rosse and Coombs (1990) found that firms placed greater emphasis on technical rather than managerial skills when recruiting technical managers. They also found that these positions remained unfilled for longer times than did engineering and scientist positions. They posit that while there were approximately equal applicants for each type of position, either many of the applicants for managerial positions were unqualified or the selection process was more protracted. Further research into the staffing practices of HT firms with respect to technical managers is needed to clarify whether these firms are using internal or external sources to fill such positions, and the level of management expertise those filling technical management positions are expected to hold.

Related to the issue of professionalization are the personal characteristics attributed to professionals. Rosenbaum (1986) lists several attributes likely to characterize professionals. One type of attribute included is mental ability, including quantitative reasoning and critical thinking. Also included, and more antithetical to effectiveness, are lack of tact, misconceptions about human behavior, and autocratic tendencies. Among the other traits listed are an emphasis on ideas over actions, a focus on organizing and understanding over dominating or persuading, and relatively low social needs. Rosenbaum also posits changing success characteristics over the span of a career, with a shift from technical competence early to a need for interpersonal competence in the longer term.

Anderson and Kleingartner (1987) portray technical professionals as high in needs for autonomy and independence, and responsive to change and uncertainty. Von Glinow (1988) describes six professional characteristics consisting of expertise, ethics, collegiality, autonomy, commitment to calling, and identification with external referents. Galbraith (1982) characterizes innovators, a reasonable proxy for the type of high-technology professional we are discussing, as high in need for achievement and as risk takers with an irreverence for the status quo. The profile we start to see is one of a highly individualistic, intellectual risk-taker with a somewhat "prickly" personality. Surprisingly, despite the apparent significance of these characteristics for the work role, little has been done to either tie them to performance or to integrate such knowledge into the selection process.

A key situational characteristic interacting with the professional personality is the rapidity of technical obsolescence experienced by the technical specialist.

It has been observed that for information workers, engineers and scientists, the pace of technological development gives the new graduate a half-life as short as five years (Rosen & Jerdee, 1985). Maidique and Hayes (1984) estimate the half-life of a Ph.D. in semiconductor engineering at seven years. This means that within five to seven years one half of the technical specialist's knowledge is no longer current. This has far reaching implications for human resource planning for staffing both in terms of recruitment of those with the most current knowledge and selection for longevity, in addition to the more obvious needs for continuing development efforts.

RECRUITMENT

In a 1985 survey of 33 Silicon Valley executives, Kleingartner and Mason (1986) found that 75% ranked recuiting as first or second in importance to HT human resource management. Their survey also showed that such firms perceive themselves as competing with a peer group of firms to hire the best and brightest technical personnel. In this sample they also found that HR specialists played only a minor role in the recruiting. Firms instead relied on close relationships with universities and the use of current technical professionals to assist with recruiting at their alma maters. The executives also indicated that they did not "raid" competitors for employees nor did they have a high concern for raiding of their employees, the feeling being that if their employees could so easily be lured away, it was the firm's own fault.

Parden (1981), in examining the mobility of technical specialists in the Santa Clara Valley, indicates that the most tangible factor in a firm's ability to recruit is salary, but that key intangibles include advancement opportunities, growth and professional development, and challenging work. Local economic conditions can help or hinder national recruitment efforts. For firms in the tight labor market and high cost-of-living Santa Clara Valley, the choice often comes down to paying a high premium to relocate someone to the area, or paying a slightly lower premium to raid other local employers.

Using interviews of HR Directors in 20 medium-sized HT firms in Philadelphia, Grissom and Lombardo (1985) found that staffing concerns differed with the growth status of the firm. Those in growth phases were most concerned with recruitment while those that were either stable or retrenching had a greater concern for motivation and productivity. Overall, recruitment and retention of technical staff were of greater concern to HT firms than to those in more traditional industries. In high-technology firms, emphasis is placed on recruitment over retention despite the high costs associated with turnover. Grissom and Lombardo conclude that there are greater similarities between stable or shrinking HT firms and traditional industries in terms of

HRM priorities. However, there is a lack of research that addresses how these priorities are reflected in the choice of recruiting techniques.

Research by Taylor and Bergmann (1987) into the effects on applicants of different recruitment activities indicates that job attributes versus "extrinsic" factors have the strongest influence on applicants in the later stages of the recruitment process. The focus on cutting edge work suggested by Von Glinow (1988) would certainly be a key job attribute relevant to HT professionals. Additionally, attributes such as advancement, growth and professional development opportunities (Parden, 1981), challenge, recognition, responsibility, and project control (Miljus & Smith, 1987) would seem important to attracting top quality personnel. These seem to parallel aspects of the professional personality described above, with appeals to achievement orientation, independence, and autonomy needs. Research demonstrating the efficacy of such appeals to the recruitment of desirable technical professionals should be a high priority.

Von Glinow (1988) indicates the importance of the literature used to attract the HT professional. It must highlight such issues as cutting edge work, good location, rewards and benefits, career advancement opportunities, and the reputation of both the firm and the key players with whom candidates may be working. She also notes that for top end technical personnel the use of search firms is the norm and the emphasis is placed on the autonomy the senior scientist will have in the position. Other individuals will respond to differing recruitment messages according to their career stage and background, with scientists and engineers differing in their orientations to the work and job attributes.

Miller (1986) also voices the importance to recruiting of communicating the nature and mission of the firm, its unique characteristics, culture, opportunities, and the high profile coworkers the prospective employee might encounter. Similar factors are cited by Miljus and Smith (1987) as well. As Turbin and Rosse (1990) point out, however, these may be the ante needed just to get into the recruiting game.

In an earlier volume in this series, Turbin and Rosse (1990) describe their review of the literature and the results of interviews they conducted with HR managers in several HT firms. They describe several barriers to recruitment that exist in HT, such as the uniformly interesting nature of the work among the firms competing for candidates, comparable compensation packages, locations which are usually very attractive but experiencing accelerated increases in cost of living, and high levels of uncertainty as to job security, especially for start-ups. They point out that these characteristics are relatively uniform across high-technology firms and thus individual firms must somehow differentiate themselves to the candidate. Based on this, they offer suggestions as to the types of practices that would appear to fit the HT firm's needs for attraction, retention, and motivation. It is for attraction that the fewest prescriptions are offered; they propose a research agenda which might shed more light on the area by evaluating the impacts of current practices and testing recommendations for improvement.

In pursuit of part of that research agenda, Rosse and Coombs (1990) found that larger firms preferentially recruit from recent college graduates, while smaller firms and those in highly competitive local labor markets favored raiding employees from other local firms. As noted earlier, Kleingartner and Mason (1985b) found that their sample of Silicon Valley executives denied raiding other firms for employees. This disagreement might reflect differences in samples, in interpretation of the term "raiding," or might result from social desirability effects. In any case larger scale studies of the external labor markets of HT firms are needed to clarify where technical professionals are being recruited and by what means.

As several articles have noted, an important recruitment strategy for HT firms has been to establish close relationships with select universities (Hill & Naroff, 1984; Kleingartner & Mason, 1986; Rozen, 1985). This might include career nights (Von Glinow, 1988), funding scholarships or subsidizing courses and programs (Cascio, 1988; Rozen, 1985), or supporting university research activities (Cascio, 1988; Kleingartner & Mason, 1986; Sparrow & Pettigrew, 1988). Locating near high quality schools that can be a source of technical personnel, as well as a source of consulting expertise and continuing education opportunities, becomes a near necessity (Hill & Naroff, 1984). A secondary consequence of the competitive labor market for scientists and engineers has been a decline in faculty in the key specialties as the best and brightest opt for jobs in industry over academia, threatening future supplies (Griggs & Manring, 1985; Solmon & La Porte, 1987). High technology firms then need dual goals of recruiting qualified personnel while increasing both the number of students going into engineering and related fields and the number of qualified faculty available to prepare future graduates (Pister, 1987; Rozen, 1985).

Cascio (1988) thinks that firms need to take a more international view of the recruitment process in HT. In his view, the labor market for HT has moved far beyond national boundaries both in terms of diversity in the labor market and the movement of operations off shore. He also focuses on the strategy of locking in a supply of technical personnel by forging tight couplings with universities through various financial supports to programs, research, and students. Characteristics of both the work itself and the environmental and professional perquisites of the job need to be emphasized in the recruiting process. He further proposes that realistic job previews that demonstrate the high levels of autonomy, responsibility, challenge, and control offered in the job are essential to effective recruitment.

Gomez-Mejia, Balkin, and Welbourne (1988) examined the role of venture capitalists in influencing HRM practices among HT firms they fund. They found agreement between venture capitalists and CEOs that the venture capitalists contribute to the recruiting of technical and managerial personnel. The venture capitalists act as information clearing houses, maintaining resumes on file, keeping track of which businesses are laying off and which are most susceptible to raiding, and making referrals to professional recruiters. Venture capitalist influences on HR practices is an underappreciated factor in studies of HT operations in the start-up through initial public offering stages of development.

While not directly applied to high-technology, Olian and Rynes (1984) attempt to integrate staffing practices with strategies described by the Miles and Snow (1978) typology. To the extent that Prospectors can be viewed as proxies of HT firms (Bretz & Dreher, 1988), the predictions made by Olian and Rynes may be applicable to the HT arena. They posit that in recruiting, Prospector type firms rely on external sources and will maximize the mix of recruiting methods used. They further predict that firms in unstable environments, as HT firms are, will use methods which provide some level of prescreening (i.e., outside agents, college placement offices, employment services) because they find it difficult to articulate stable hiring specifications or procedures. Their recruitment channels may be less formal and operate through line personnel rather than through HR specialists. The Rosse and Coombs data offer mixed insights into these propositions. Referrals were indicated as an important technique, though the use of headhunters and on-campus interviewing ranked as only minor factors. As will be noted again in the discussion of selection techniques, line personnel played a greater role than do the HR staff, supporting the Olian and Rynes prediction.

Some HT firms have been associated with a trend to hire retirees (Perham, 1982). The House Select Committee on Aging identified three industries—high-technology, fast food, and insurance—as being active in bringing retirees back into the workforce. The motivation in HT is the shortage of qualified technical personnel, and such firms as Grumman and Lockheed have established special programs to employ retired engineers and technical personnel. It would seem that such a source is most available to older firms that have a pool of retirees to tap and those with relatively stable technologies. Start-ups and other young firms would seem less able to find the type of personnel likely to be immediately productive or with the current technical knowledge necessary to participate in cutting edge projects. With the graying of the workforce and declining domestic enrollments in the sciences, however, this may be a recruiting source with increasing importance and expanding applicability in the future.

SELECTION

Gattiker (1988), in providing some guidelines for future research into the management of technological development, makes the following observation:

> In order to recruit individuals who closely fit the organization and its culture, it is necessary to improve personnel selection and testing. Trial and error is much too costly, as turnover becomes ever more expensive for companies. An extensive human resource accounting system ought to be in place to allow management to quantify the cost of errors in recruiting. Thus the selection instruments must be validated before they can be used effectively. (p. 216)

However, Gattiker does not provide any prescriptions for improving selection. Neither does Zimmerman (1986), who suggests that firms make selections more meaningful by identifying indicators that predict success and cultural fit yet offers no guidance as to the means to do so.

Bretz and Dreher (1988) take a more prescriptive approach. Applying Staw's (1986) framework for grouping HR practices into logical systems and matching these against various strategic typologies, they attempt to describe combinations which seem to fit the characteristics of high technology. In this attempt they find an apparent conflict. Due to the high reliance on technical professionals, an individually-oriented system would be preferred. Emphasis should be placed on individual ability with recruitment and selection focusing on past achievement (though given the short half-life of knowledge, this seems somewhat counterintuitive) and achievement needs. On the other hand, the strategies associated with HT firms call for an organizationally-oriented system that should preferentially select for aptitude and future contribution. Bretz and Dreher suggest that an attempt be made to selectively introduce characteristics of an individually-oriented system, while retaining the longer-term focus of the organizational orientation. Specific recommendations for techniques to implement this are, unfortunately, not provided.

Galbraith (1982) focuses on the differing characteristics to select for in choosing technical personnel and technical managers for an innovation-oriented organization. The innovators match the profile of the technical professional described earlier and the managers must be able to deal with these difficult traits. Galbraith mentions negotiation and brokering skills as key traits of the effective technical manager and favors generalists with strong business skills and experience with innovation. Schuler and Jackson (1989) similarly consider innovation as a competitive strategy and how human resource practices support strategic thrusts. They focus on the employee behaviors needed for innovation and the importance of implementing HR practices to reinforce the desired behaviors. This implies that both recruitment and selection need to be tailored to the specific type of employees needed to pursue an innovation strategy, though they do not make specific recommendations. Schuler (1987) uses the Gerstein and Reisman (1983) typology of corporate strategies and ties general recommendations to each type. High-technology firms would seem to be primarily Entrepreneurial or Dynamic Growth firms in the Gerstein and Reisman system, depending upon their stage in the life cycle. In both cases the recommendations made for staffing include broad paths, implicit criteria, and openness in the system. Once again, recommendations are not made as to the specific types of selection practices which might result from these general guidelines.

A role-based approach is offered by Roberts and Fusfeld (1981). They identify five informal, critical behavioral functions or roles needed to effectively generate technological innovations. As each role is carried out by relatively few individuals in the organization, replacement when someone leaves is difficult and costly. Traditionally, personnel are selected for their technical skills rather than their abilities to fill these more general roles, often leaving gaps in critical innovation functions. The authors posit that specific personal characteristics predispose individuals to particular roles and each of these types needs to be recruited and selected differently. In their view, all technical personnel are hired and managed as if they are either highly creative scientists or routine technical problem solvers, with little regard given to the mix of skills needed for effective innovation. They suggest that the characteristic strengths of the roles be translated into needed skill sets and these skills be used as the basis of interviewing.

Heyer (1985) offers a dynamic skill flow model to assist with human resource planning. His model tracks additions, transitions, retraining, surpluses, and losses from the personnel system by personal characteristics and skills to allow planning for specific skill needs within a five-year horizon. One difficulty with applying his model is that he assumes a relatively stable organization with employment security, career paths, internal promotions, retraining and dual career ladders. While this is applicable to a subset of high-technology firms, it will not be generally applicable.

Empirical studies of selection practices in use by HT firms are rare. Rosse and Coombs (1990) surveyed Directors of Personnel and found the most commonly used selection techniques to be very mundane: interviews by line managers, reference checks, referrals from employees or college professors, and interviews by HR staff. Despite the conventional wisdom that HT firms are likely to use progressive human resource practices, Rosse and Coombs found little evidence of such practices as weighted application blanks, realistic job previews, or ability testing in use for the selection of scientists, engineers, or technical managers. In light of this, it is difficult to believe that HT firms are achieving the level of success they might in selecting personnel with the appropriate mix of characteristics needed.

The heavy use of interviews is especially interesting in that it reflects a dilemma for technical fields. Line personnel have the acumen to probe the candidate's technical background and knowledge but may lack the interpersonal skills to make a good impression on candidates and the HR knowledge to conduct consistent and valid interviews. The trained HR specialist may be able to conduct valid interviews but may lack the technical know-how to effectively evaluate the candidate's potential contribution. This suggests that the optimal role for the HR professional may be to act as a consultant in the development of interview protocols that are administered by technical specialists. Moreover, the interview process may be best suited to determining the applicant's fit with the organization's culture, while other techniques, such as technical presentations by applicants and panel interviews by technical specialists may provide better insight into the research potential of applicants (Wolff, 1984). Bleicher, et al. (1983) suggest coworker interviews to develop a self perpetuating work-group culture. Von Glinow (1988) proposes that the selection process must be in the hands of the project manager rather than the HR staff. She also indicates the importance of choosing the correct technical interviewer to best represent the firm. Again, empirical research into the efficacy of such techniques in the HT setting is lacking.

Olian and Rynes (1984) suggest that Prospector firms are likely to have selection standards which are in a constant state of revision. The types of selection techniques they feel match this strategic thrust include informal exchanges between applicants and representatives rather than formal screening; emphasis on evaluating work history through ability testing and reference checks instead of aptitude testing to determine potential; methods emphasizing verbal skills; and a strong line influence on hiring decisions. The results reported by Rosse and Coombs (1990) support some of these contentions for the selection of scientists and engineers in HT, with heavy reliance on interviews by line managers and reference checking. Skill assessments are used more commonly than either specific or general ability tests, though none of these are used with great frequency.

In a study of nine companies in the computing systems industry, Sparrow and Pettigrew (1988) found that the firms were reorienting their HR practices in response to new environmental trends toward slower growth and declining profitability. One shift was toward selecting graduates with business knowledge and experience in addition to technical skills. This would seem to narrow the field substantially and increase the difficulty in finding qualified prospects in a tight labor market. It would be of interest to know more about what these firms had been doing in the past and to determine if such a trend is true more generally.

In a case study of an electronics firm, Perry and Sandholtz (1988) examined an extreme case of a firm making radical product innovations. They suggest that such an organization requires a "liberating form" characterized by open design and professional values supportive of improvisation. The types of values found are described as academic, emphasizing research, design and development, a technological imperative, and technical arrogance. Perry and Sandholtz apply the concept of the utilitarian organization to suggest that such an organization would delegate socialization of employees to professional schools and focus recruitment and selection on referrals and carefully selected universities which provide the desired type of socialization. They suggest marathon interviews, showing candidates work-in-progress, and self-selection as keys to effective staffing. They also indicate that such a system encourages high turnover, comparable to a rigorous university tenure system. While few firms are likely to pursue as radical an approach as Perry and Sandholtz describe, it is intriguing to look at their description as one end of a continuum of possibilities for the HT firm.

CONCLUSIONS AND RECOMMENDATIONS FOR FUTURE RESEARCH

There is a conflict between assertions that HT firms use innovative or progressive HR techniques and the results reported in the few empirical studies found. The first priority of a research agenda for staffing issues in high-technology needs to be one or more large sample surveys of techniques in use by high-technology firms, with a focus on the effectiveness of the techniques utilized.

A key issue which needs to be addressed, not only in studies of staffing practices but more generally in the study of high-technology industry is whether or not there are identifiable differences in strategy among HT firms. It is readily accepted that HT firms are similar to Prospectors or follow innovation strategies, but might there be variants among these industries that require a finer grained examination? Jackson, Schuler, and Rivero (1989) consider a series of organizational characteristics which predict personnel practices. Empirical evaluation of HT firms along these or similar dimensions might open up new horizons for understanding such companies. Another approach would be to test the propositions offered in many of the papers described above, especially with respect to the match between corporate strategies and supportive HR practices. Kochan and Chalykoff (1987) state that HT firms should be expected to have ineffective HR policies due to the difficulty of behaving strategically in the face of rapidly changing environments, yet they suggest that HT firms have the appropriate value systems to adopt a new human resources model, at least during the early stages of their life cycles. They provide a set of propositions in need of empirical testing.

Another area requiring a finer grained approach is the technical professional. Different authors variously speak to their differences or focus on their similarities. Empirical analysis of the high-technology technical professional with an eye toward differences which may exist within the engineering and the scientist groups may turn up many interesting issues. How have HR practices diffused through high-technology industries over time? Kossek (1987) offers some propositions for the diffusion of HR innovations which could be applied in the special milieu of high-technology. Additional questions also present themselves. How do the HR practices in place change over the life cycle transitions of the HT firm? How must HR practices be adapted to the changing character of the technical professional? These and related questions offer a plethora of research opportunities.

The entire concept of professionalism among HT personnel requires greater attention. Questions need to be asked about the similarities and differences between HT professionals and other technical professionals in more traditional industries. If the industry itself is creating a self-fulfilling prophecy of professionalization, what implications does that hold for future staffing? Technical managers have also received too little attention. While there is a long standing tradition of looking at managers in R&D settings, HT seems to offer a unique set of circumstances impacting on managerial behavior and calling for special skills. Several authors have suggested what those behaviors and skills might be, but research is needed to confirm or challenge those assertions. Clarification is also needed as to how HT firms acquire technical managers, whether from internal or external labor markets, using what selection techniques, and with what managerial qualifications.

The interaction of compensation issues with other HR practices needs also to be examined. This hearkens back to the statement made by Anderson and Kleingartner as to the dearth of empirical research into the actual effects of the human resource practices in use among HT firms. There is a need for a unifying analytical framework for the field. Mahoney and Deckop (1986) suggest a focus on labor market structures, following Kerr (1954). The structures described, however, seem inadequate to describing the labor market structure of high-technology firms. How might high-technology labor markets be added to the framework?

Maidique and Hayes (1984) propose that HT requires a periodic shift between continuity and chaos as the firm progresses through its life cycle. Longitudinal studies on this question might also examine the shifts in HR practices needed to keep up with the changing needs of the organization. What types of personnel are able to tolerate both the chaos and the continuity phases? How can they be identified during the selection process?

A basic question which underlies all of the issues raised above is: Are high-technology firms really as different as we all assume? This question can be addressed in terms of HR practices, strategy, and many other important variables. Some of the problems encountered in answering this question rest on difficulties in drawing clear boundaries around high technology. Dichotomizing industries or firms into HT and not-HT fails to recognize the continuous nature of the factors used to identify technological intensity.

The lack of empirical research into recruitment and selection practices in HT firms is the most notable finding of this review of the literature. The empirical pieces identified are hampered by small sample sizes and, in some cases, highly localized samples, making it difficult to make many generalizations. It may be that research is currently in the pipeline in response to many of the prescriptions and propositions that have been offered in the theoretical and practical papers cited and future reviews may find greater clarification of research into staffing issues in high technology.

REFERENCES

Anderson, C.S., & Kleingartner, A. (1987). Human resource management in high-technology firms and the impact of professionalism. In A. Kleingartner & C.A. Anderson (Eds.), *Human resource management in high-technology firms* (pp. 2-21). Lexinton, MA: Lexington Books.

Badawy, M.K. (1971, Fall). Industrial scientists and engineers: Motivational style differences. *California Management Review*, pp. 11-16.

Badawy, M.K. (1988, September-October). What we've learned: Managing human resources. *Research Technology Management*, 19-35.

Balkin, D.B., & Gomez-Mejia, L.R. (1985, June). Compensation practices in high technology industries. *Personal Administrator*, pp. 111-118, 122-124.

Belous, R.S. (1987). High technology labor markets: Projections and policy implications. In A. Kleingartner & C.A. Anderson (Eds.), *Human resource management in high-technology firms* (pp. 25-45). Lexington, MA:Lexington Books.

Bestuzhev-Lada, I. (1987, June). High technology and long-term global problems. Futures, pp. 276-281.

Bleicher, H.A., Bleicher, F., & Paul, H. (1983, November-December). Managerial frameworks for innovative responses in high-tech organizations. *Business Horizons*, pp. 69-78.

Bretz, R.D., & Dreher, G.F. (1988). Individual, group, and organizationally oriented personnel systems: Implications for staffing the high-technology firm. *Proceedings '88: Managing the High Technology Firm*, pp. 2-5.

Cascio, W.F. (1988). Strategic human resource management in high technology industry. *Proceedings '88: Managing the High Technology Firms*, pp. 9-15.

Galbraith, J.R. (1982, Winter). Designing the innovating organization. *Organizational Dynamics*, 5-25.

Gattiker, U.E. (1988). Where do we go from here? Directions for future research. In U.E. Gattiker & L. Larwood (Eds.), *Managing technological development: Strategic and human resources issues* (pp. 213-217). Berlin: Walter de Gruyter & Company.

Gerpott, T.J., & Domsch, M. (1985). The concept of professionalism and the management of salaried technical professionals: A cross-national perspective. *Human Resource Management, 24*(2), 207-226.

Gerstein, M., & Reisman, H. (1983, Winter). Strategic selection: Matching executives to business conditions. *Sloan Management Review*, pp. 33-49.

Gomez-Mejia, L.R., Balkin, D.B., & Welbourne, T.M. (1988). The influence of venture capitalists on human resource practices in the high technology industry. *Proceedings '88: Managing the High Technology Firm*, pp. 28-34.

Gomez-Mejia, L.R., & Welbourne, T.M. (1990). The role of compensation in the human resource management strategies of high-technology firms. In M.A. Glinow & S.A. Morham (Eds.), *Managing complexity in high technology organizations* (pp. 255-277). New York: Oxford University Press.

Griggs, W.H., & Manring, S.L. (1985, June). Money isn't the best tool for motivating technical professionals. *Personnel Administrator*, pp. 63-76.

Grissom, G.R., & Lombardo, K.J. (1985, June). The role of the high-tech HR professional. *Personnel*, pp. 15-17.

Hayes, R.H., & Abernathy, W.J. (1982). Managing our way to economic decline. In M. Tushman (Ed.), *Readings in the management of innovation* (pp. 11-25). Boston, MA: Pitman.

Heyer, N.O. (1985). Managing human resources in a high-technology enterprise. In R.J. Niehaus (Ed.), *Human resource policy analysis: Organizational applications* (pp. 45-66). New York: Praeger.

Hill, J., & Naroff, J.L. (1984, Spring). The effect of location on the performance of high-technology firms. *Financial Management*, pp. 27-36.

Hybels, R.C., & Barley, S.R. (1988). Co-optation and the legitimation of professional identities: Human resource policies in high-technology firms. *Proceedings '88: Managing the High Technology Firm*, pp. 35-42.

Jackson, S. E., Schuler, R.S., & Rivero, J.C. (1989). Organizational characteristics as predictors of personnel practices. *Personnel Psychology, 42*(4), 727-786.

Kanter, R.M. (1984). Variations in managerial career structures in high technology firms: The impact of organizational characteristics on internal labor market patterns. In P. Osterman (Ed.), *Internal labor markets* (pp. 109-131). Cambridge, MA: MIT Press.

Kerr, C. (1954). The Balkanization of labor markets. In E.W. Bakke, P.M. Hauser, G.L. Palmer, C.A. Myers, D. Yoder, & C. Kerr (Eds.), *Labor mobility and economic opportunity* (pp. 93-109). New York: Wiley.

Kerr, S., & Von Glinow, M.A. (1977). Issues in the study of "professionals" in organizations: The case of scientists and engineers. *Organizational Behavior and Human Performance, 18*, 329-345.

Kleingartner, A., & Mason, R.H. (1986, April). Management of creative professionals in high-technology firms. *IRRA Proceedings*, pp. 508-515.

Kochan, T.A., & Chalykoff, J.B. (1987). Human resource management and business life cycles: Some preliminary propositions. In A. Kleingartner & C.A. Anderson (Eds.), *Human resource management in high-technology firms* (pp. 183-200). Lexington, MA: Lexington Books.

Kossek, E.E. (1987). Human resources management innovation. *Human Resource Management, 26*(1), 71-92.

Mahoney, T.A., & Deckop, J.R. (1986). Evolution of concept and practice in personnel administration/human resource management (PA/HRM). *Journal of Management, 12*(2), 223-241.

Maidique, M.A., & Hayes, R.H. (1984, Winter). The art of high-technology management. *Sloan Management Review*, pp. 17-31.

Miles, R.E., & Snow, C.C. (1978). *Organizational strategy, structure, and process*. New York: McGraw-Hill.

Miljus, R.C., & Smith, R.L. (1987). Key human resource issues for management in high-tech firms. In A. Kleingartner & C.A. Anderson (Eds.), *Human resource management in high-technology firms* (pp. 115-131). Lexington, MA: Lexington Books.

Milkovich, G.T. (1987). Compensation systems in high-technology companies. In A. Kleingartner & C.A. Anderson (Eds.), *Human resource management in high-technology firms* (pp. 103-114). Lexington, MA: Lexington Books.

Miller, D.B. (1986). *Managing professionals in research and development*. San Francisco: Jossey-Bass.

Olian, J.D., & Rynes, S.L. (1984). Organizational staffing: Integrating practice with strategy. *Industrial Relations, 23*(2), 170-183.

Parden, R.J. (1981). The manager's role and the high mobility of technical specialists in the Santa Clara valley. *IEEE Transactions on Engineering Management, EM-28*(1), 2-7.

Perham, J. (1982, March). Putting retirees to work. *Dun's Business Month*, pp. 80-82.

Perry, L.T., & Sandholtz, K.W. (1988). A "liberating form" for radical product innovation. In U.E. Gattiker & L. Larwood (Eds.), *Managing technological development: Strategic and human resources issues* (pp. 9-31), Berlin: Walter de Gruyter.

Pister, K.S. (1987). The dynamic relationship between research universities and the development of high-technology industry. In A. Kleingartner & C.A. Anderson (Eds.), *Human resource management in high-technology firms* (pp. 67-81). Lexington, MA: Lexington Books.

Roberts, E.B., & Fusfeld, A.R. (1981, Spring). Staffing the innovative technology based organization. *Sloan Management Review,* pp. 19-34.

Rosen, B., & Jerdee, T.H. (1985, March). A model program for combating employee obsolescence. *Personnel Administrator,* pp. 86-92.

Rosenbaum, B.L. (1986, October). Critical skills for successful technical professionals. *Personnel,* 56-60.

Rosse, J.G., & Coombs, G. (1990). *In search of the elusive technical specialist: A survey of hiring practices among high-tech firms.* Working paper, University of Colorado Graduate School of Business and Administration, Boulder.

Rozen, M. (1985, March). Wanted: High-tech engineers. Dun's Business Month, 35-36.

Schuler, R.S. (1987). Personnel and human resource management choices and organizational strategy. *Human Resource Planning, 10*(1), 1-17.

Schuler, R.S., & Jackson, S.E. (1989). *Linking compensation practices to innovation as a competitice strategy.* Working paper, New York University, New York.

Sherman, J.D. (1989). Technical supervision and turnover among engineers and technicians: Influencing factors in the work environment. *Group & Organization Studies, 14*(4), 411-421.

Solmon, L.C., & La Porte, M.A. (1987). The educational implications of the high-technology revolution. In A. Kleingartner & C.A. Anderson (Eds.), *Human resource management in high-technology firms* (pp. 47-66). Lexington, MA: Lexington Books.

Sparrow, P., & Pettigrew, A. (1988, February). Contrasting hrm responses in the changing world of computing. *Personnel Management,* pp. 40-45.

Staw, B. (1986). Organizational psychology and the pursuit of the happy/ productive worker. *California Management Review, 28*(4), 40-53.

Taylor, M.S., & Bergmann, T.J. (1987). Organizational recruitment activities and applicants' reactions at different stages of the recruitment process. *Personnel Psychology, 40,* 261-285.

Turbin, M.S., & Rosse, J.G. (1990). Staffing issues in the high technology industry. In L.R. Gomez-Mejia & M.W. Lawless (Eds.), *Organizational issues in high-technology management* (pp. 227-241). Greenwich, CT: JAI Press.

Von Glinow, M.A. (1988). *The new professionals: Managing today's high-tech employees.* Cambridge, MA: Ballinger.

Wolff, M.F. (1984, January/February). Hiring people who do good research. *Research Management,* pp. 8-9.

Zimmerman, J.H. (1986, April). Human resource management at MCI. *Management Review,* pp. 49-51.

INNOVATORS AND NONINNOVATORS: A COMPARISON OF DEMOGRAPHIC, PSYCHOLOGICAL, BEHAVIORAL, AND ROLE FACTORS

Dennis Duchon, Donde P. Ashmos, and Kenneth J. Dunegan

INTRODUCTION

All human resource functions, even in high technology firms, have some responsibility for four generic activities: selection or recruitment of qualified personnel, evaluation, rewards, or maintaining adequate compensation and benefits packages; and development or creating systems which enhance skills, promotional opportunities, and career paths (Fombrun & Tichy, 1984). In this volume Coombs and Rosse (this volume, pp. 91-107) discuss issues surrounding the recruitment of professionals. Katz, Tushman, and Allen (this volume, pp. 133-150) address reward and development issues through an examination of the dual

Human Resource Strategy in High Technology.
Advances in Global High-Technology Management, Volume 1, pages 109-123.

ISBN: 1-55938-489-1

ladder; and Drory (this volume, pp. 125-132) examines how reward, evaluation, and development issues impact the intention to leave an organization. The present paper examines an approach to selection and development by looking at a special subset of professionals: people identified as *innovators*. These innovators are compared to *noninnovators* in terms of demographic, psychological, behavioral, and role factors. The concept of *risk taker* is used as a unifying theme, and the paper challenges the assumption that innovators are risk takers.

BACKGROUND

Why It is Important to Identify Innovators

In both the popular and academic literatures it is impossible to get away from the idea of innovation. Because of domestic and international competition, changing government regulations, changing market conditions and so on, organizations must be innovative in order to survive (Amabile, 1988). In a larger sense the accomplishment of innovation rests on an organization's willingness and capacity to change because innovation is, by definition, a significant departure from current practice.

Yet, a great body of research has demonstrated that organizations are not readily subject to change and, by implication, not responsive to innovation efforts either (cf. Hannan & Freeman, 1977, 1984; McKelvey, 1982; Meyer & Rowan, 1977; Romanelli, 1989). Kanter (1988), for example, argues that patterns in most well-established organizations inhibit rather than activate innovation. Quinn (1985) contends that these inhibiting forces are the result of excessive rationalism and bureaucracy.

What, then, is an organization to do if it needs to be innovative but recognizes that its systems will resist innovation? The organization tries to identify people, individuals, who are able to ignore or resist the forces of excessive rationalism and bureaucracy. Coombs and Rosse (this volume, pp. 91-107) argue that high-tech organizations live or die by the quality of the people they hire. Applying the Coombs and Rosse notion to the realm of innovation, the belief, or hope, is that special *individuals,* because of their skills and energies, will bring about innovation in spite of the pressures for conformity.

Organizations, in fact, act on this belief in the power of individuals to achieve innovation by using innovative or creative potential as a hiring criterion. Hewlett-Packard, for example, is building a $40 million "factory of the future" in Puerto Rico where computer systems employees will be hired on the basis of their creative potential (Stead, 1990). Thus, the identification of people who are believed to be able to engage successfully in innovation is already an active part of human resource management.

Innovators and Their Roles

In general, innovators are seen as those people in an organization who are the dreamers (Pinchot, 1988), the creative ones, and, importantly, the risk takers. Moreover, these entrepreneurial spirited employees are believed to differ from their noninnovative counterparts in their personalities and their behaviors (cf. Knight, 1967; Schrage, 1965).

Innovators in contrast to non-innovators are in general argued to be more likely to register independence of judgment, will deviate more from norms, will be more impulsive, and will exhibit a preference for conceptual flexibility and complex tasks (Steiner, 1965). Innovators will be curious, persistent, self motivated, and energetic (Amabile, 1988). They may be troublemakers (Quinn, 1985).

But simply identifying people who are thought to possess certain characteristics or capabilities is not the end product of effective human resource management because these people must be matched to *roles* in the organization. There must be an innovator role for them to fulfill. Romanelli (1989), for example, argues that traditionally an organization's potential for innovative activity is viewed either in terms of the characteristics of the innovators or in terms of the conditions in the organization which create opportunities for innovation.

Ultimate innovative activity, however, does not depend on either the person or the organizational environment, but rather the interplay of both (Amabile, 1988). Innovation requires a person to capitalize on conditions, someone to take action given the conditions. Thus, it is important not only to select the innovator, but also to match him/her to a role in the organization. Later it will be important to develop systematically the person in the role.

The role of innovator will be different depending on the size and complexity of the organization involved (Hisrich, 1990). In the classic small entrepreneurial organization the innovator plays many roles (e.g., opportunity identifier, resource manager, fund raiser, and so forth). In the large, complex organization the roles typically are narrowly defined and assumed by different actors.

Kanter (1988), for example, discusses four tasks (roles) attending innovation activity: idea generation activity, coalition building and power acquisition, idea realization, and transfer or diffusion. Roberts and Fusfeld (1988) expand the notion of role differentiation in the innovation process by suggesting five roles: idea generator, entrepreneur/product champion, project leader, gatekeeper, and sponsor.

An important implication of differentiating roles in the innovation process is the likelihood that no one individual is well suited to all of these roles. Roberts and Fusfeld (1988), in fact, argue that specific personal characteristics predispose individuals to particular roles, and each of these "types" (roles) needs to be recruited and selected differently.

Innovator as Risk Taker

Innovation requires a step into the unknown, and such a step carries with it a potential for reward and gain, but it also carries a potential for loss. MacCrimmon and Wehrung (1986) in fact define a risky situation as one in which the outcome is uncertain and there exists a potential for loss. Risk takers are those who are willing or able to operate in a risky environment.

A great deal has been written on the identity of innovators as risk takers, and in the popular management literature innovators are simply assumed to be risk takers (cf. Peters & Austin, 1985; Waterman, 1988). In the academic literature, however, the connection between the role of innovator and the label "risk taker" is not always so clear. Some researchers state unequivocally that innovators are risk takers (cf. Amabile, 1988; Galbraith, 1984; Steiner, 1988). Other researchers claim that innovators take moderate, calculated risks (cf. Kets de Vries, 1985; Low & MacMillan, 1988). Yet others argue that risk taking is part of some of the roles of innovation (e.g., idea generator, project champion), but not necessarily all the roles (cf. Roberts & Fusfeld, 1988). And still others claim that risk taking propensity may not even be a distinguishing characteristic of innovators (cf. Brockhaus, 1980).

MacCrimmon and Wehrung (1986) bring some coherence to the diffused discussion of innovator as risk taker by noting that researchers have not achieved a standard way to assess risk. Whether an innovator (or anyone else) is seen as an extreme, moderate, or low risk taker depends on the method used to assess risk. Risk taking is also situational. People may be risk takers in one situation, but not another. A physical risk taker like a sky diver may not be a financial risk taker like someone who buys penny stocks. A financial risk taker may not take on the career risks of battling the bureaucracy.

Thus, despite a popular belief in the "innovator as risk taker" the empirical evidence suggests that while there is an innovator role(s), there may be no such creature as a transcendent risk taker. If there is no such thing as a universal risk taker type, then, in a human resource management sense, it is a vain hope to select innovators simply because they appear to be risk takers.

This study compares demographic features, psychological characteristics, behaviors, and job (role) perceptions of two groups: one group believed by their managers to be innovators and a second group believed to be noninnovators. The expectation is that while the groups will differ in some ways, no clear risk taking "type" will emerge.

METHOD

Procedure

This study was conducted in a Fortune 500 chemical company. The company is international in its operations and multidivisional in its structure. Questionnaires assessing individual difference and organizational structure variables were administered to every scientist, engineer, and manager in one division of the company ($n = 316$). These questionnaires were returned directly to the research team to ensure respondent confidentiality ($n = 239$, response rate = 76%).

Sample

Top executives representing the manufacturing, R&D, and corporate management functions of the division identified individuals they felt were either innovative or not innovative. After some discussion executives considered an innovative person as one who was capable of producing a product or process, or an idea leading to a product or process that is new to the organization (Zaltman, Duncan, & Holbek, 1973). The definition of innovator used in the study thus reflects the popular view of who is an innovator and it focuses on the role Roberts and Fusfeld (1988) call *idea generator.*

The executives identified 46 individuals who had completed questionnaires. The nomination process thus resulted in a sample of 22 innovators and 24 noninnovators.

Members of the sample were considered by the executives to be mid-level managers. On average they occupied level 6 in an 11-level hierarchy. To ensure that the innovator-noninnovator nomination process was not solely a function of job performance, the two groups were compared on the basis of their company performance ratings. The innovator group had an average rating of 7.4 (on a 10-point scale) and the Noninnovator group an average rating of 6.8. This difference is not statistically significant ($t = 1.54$, ns). The remainder of the sample ($n = 193$) on average occupied level 7 in the hierarchy and had a company performance rating of 6.6.

Measures of Risk Taking

This study compares innovators and noninnovators in terms of demographic, psychological, behavioral, and role perception factors which might indicate risk taking.

Demographic Indicators

Demographic comparisons are listed in Table 1. Note that the demographics of the remainder of the sample (referred to as "Others") are also listed.

MacCrimmon and Wehrung (1986) have offered evidence that younger employees, employees who have been with the firm a short time, and more highly educated employees are risk takers.

Psychological Indicators

The psychological indicators of risk taking included two personality scales and one self-assessment measure. One of the personality measures was an adaptation of the Kirton Adaption-Innovation Inventory (Kirton, 1976). The KAI is a measure of a person's inclination to "do things better" versus an inclination to "do things differently." People who do things "better" are called Adaptors and are characterized by precision, reliability, and efficiency. They tend to be concerned about resolving problems, and they seek problem solutions by using tried and true methods. They tend to provide stability, order, and continuity to their work groups. In the context of the present study Adaptors would be expected to be risk avoiders.

People who do things "differently" are called innovators and are characterized as tangential thinkers who may approach tasks from unexpected angles. They tend to find problems then challenge the assumptions underlying those problems. Innovators can help their work groups break with the past. In the context of the present study, innovators would be expected to be risk takers.

The original KAI (Kirton, 1976) was composed of 32 items. Based on psychometric work conducted by Keller and Holland (1978a, 1978b), the present study used a 13-item version of the original instrument. The instrument requires subjects to imagine presenting a certain image of themselves to others. They are then asked to rate the *difficulty* of presenting such an image for each of the traits or attributes stated in the 13 items. Responses to the items are indicated on a 5-point scale which is anchored by "very easy" and "very difficult." The scale is scored so that innovators are higher than the mean and adaptors are lower than the mean. Cronbach's alpha for the scale in the present study was .86. The 13-item scale is reproduced in Appendix A.

Keller and Holland (1978a, 1978b, 1979) have noted the utility of using the KAI for describing style differences in decision makers, for studying creativity, or predicting employee "fit" in innovative tasks. The innovators and noninnovators nominated for the present study would be expected to have different KAI scores.

In addition to the KAI, subjects' risk taking propensity was assessed using a modification of a risk taking scale on the *Jackson Personality Inventory Manual* (Jackson, 1976). People scoring high on this scale are expected to enjoy taking chances, be willing to expose themselves to situations with uncertain outcomes, and enjoy adventures with an element of peril. Low scorers, or risk avoiders, on the other hand, are expected to be cautious in unpredictable situations, and even avoid situations of personal risk even those offering great rewards.

In the original *Jackson Personality Inventory Manual,* subjects are asked to agree or disagree with an item. In the present study the scale was modified. Subjects indicated their response to an item on a 5-point scale that was anchored by 1 = not at all and 5 = very great extent. The items used were: (1) I try to avoid situations that have uncertain outcomes (reverse scored); (2) I rarely, if ever, take risks when there is an alternative (reverse scored); (3) When I want something, I'll go out on a limb; (4) People think that I enjoy taking chances; (5) I would enjoy the challenge of a project that could mean either a promotion or a loss of a job. Cronbach's alpha for this measure was .62 in the present study.

Jackson has provided evidence that his measure will differentiate risk takers and risk avoiders (Jackson, 1976, 1977). In the present study we would expect the nominated Innovators to have higher risk taking scores than the nominated Noninnovators.

Finally, the third personality measure of risk taking was a self assessment. Subjects were asked to respond to the following single item: "In general I consider myself to be." Subjects placed themselves on a scale which was anchored by 1 = a risk avoider and 5 = A risk taker. MacCrimmon and Wehrung (1986) have noted that people are not always aware of their own risk taking propensities, however higher level managers tend not only to see themselves as risk takers but also prove to be risk takers. In the present study we would expect the nominated Innovators to see themselves more as risk takers than the nominated Noninnovators.

Behavioral Indicators

Subjects were asked to read a scenario and assume the role of a manager who faces a risky decision. The scenario is reproduced in Appendix B. After reading the scenario subjects responded to three questions:

1. What is the likelihood you would fund the project? (1 = reject, 2 = lean toward rejecting, 3 = uncertain, 4 = lean toward funding, 5 = fund);
2. Indicate the level of risk you believe is associated with providing the additional funding (1 = no risk, 2 = some risk, 3 = normal risk, 4 = considerable risk, 5 = too risky);
3. What probability of success for the project would be necessary in order for you to fund it? (____% chance of success).

The executives who nominated the innovator and noninnovator groups judged the scenario to be a reasonable depiction of the kinds of decisions made in the organization.

We would expect in the present study that the nominated Innovators are more inclined to fund the project, see less risk in the situation (that is, risk takers do not see situations as risky that is why they "take chances"), and require a lower probability of success than nominated noninnovators.

Role Indicators

Separate from the characteristics of an individual who might be seen as an innovator ("risk taker") is the role the individual fulfills in an organization. Role or task demands have been demonstrated to be a potent factor in eliciting risk taking behavior (Dunegan & Duchon, 1989). Therefore people might be seen by their superiors to be risk takers because of role requirements. Two role indicators were assessed in this study, both of them perceptual because it is the role incumbent's perception of and interpretation of his/her role that will lead ultimately to behavioral differences on the job.

One role indicator pertinent to risk taking is Perrow's (1967) view of task uncertainty which is assessed by measuring task analyzability and task variety. Analyzability refers to the extent a person can follow unambiguous processes to solve task related problems, and variety refers to the number of exceptions, unexpected, and/or novel events found in the job. As a role or job is characterized as unanalyzable and has much variety then it is said to be uncertain. Innovators (risk takers) would be expected not only to perform better in uncertain circumstances, but also they are likely to perceive their work environment as uncertain so that they might thus define for themselves a psychological work space which allows innovative activity.

The measures used in the present study were adapted from Withey, Daft, and Cooper (1983). Analyzability and variety both were measured using a 5-point Likert type scale anchored by 1 = to a small extent, 3 = to some extent, 5 = to a great extent.

The questions assessing Analyzability (Alpha = .79) included:

- Is there a clearly known way to do the major types of work you normally encounter?
- Is there a clearly defined body of knowledge of subject matter which can guide you in doing your work?
- Can you actually rely on established procedures and practices as do your job?
- Do you follow standard operating procedures and practices as you do your job?
- Is there an understandable sequence of steps that can be followed in carrying out your work?

The questions assessing variety (Alpha = .73) included:

- Do people in this department do about the same job in the same way most of the time?
- Basically, department members perform repetitive activities in doing their jobs
- Are your duties repetitious?

Analyzability and variety scores are reported in Table 2.

A second role indicator required subjects to describe their jobs in terms of independent, sequenced, or interdependent events. This measure was based on Thompson's (1967) conceptualization of interdependencies in the workplace. Popular views typically cast the innovator in the role of loner: an autonomous, independent agent. In contrast, however, the nature of modern organizational life with its complex networks of roles suggests that innovators see themselves not as loners but as members of interdependent relationships. Thus it will be interesting to see if innovators in a large organization see their work roles as independent or interdependent.

In order to measure job dependency subjects were told that work assignments could be described as:

Type I. Independent, work assignments are done completely by one person.
Type II. Sequenced, one person's work follows the work of another person.
Type III. Interdependent, work assignments involve many people at the same time.

Subjects were then asked: What percentage of your work fits the description of: Type I? Type II? Type III? Percentages for innovators and noninnovators are reported in Table 2.

RESULTS

The data in Table 1 suggest that the innovator group is younger, has been with the firm a shorter period of time, and occupied their present positions a shorter period of time than the noninnovator group. The innovator group is, on average, better educated than the noninnovator group. None of these differences, however, reach conventional levels of statistical significance. One difference that is statistically significant is gender. There is a larger number of females in the innovator group than the noninnovator group (Chi-Square $= 4.71, p < .05$).

The data in Table 2 show that the innovator group can be considered to be risk takers according to some, but not all, of the measures employed in the study. All of the psychological measures reveal significantly different scores for innovators and noninnovators. Moreover, these scores are in the expected direction; that is, innovators are risk takers. According to the KAI measure, innovators get innovation scores and noninnovators get adaptor scores (mean 53.7 vs. 43.8, $F = 25.81, p < .001$). Innovators get higher Jackson risk taking scores (mean 3.5 vs. 2.9, $F = 9.69, p < .01$). Innovators also report themselves to be more risk taking than noninnovators (mean 3.6 vs. 3.0, $F = 6.74$, $p < .01$).

Table 1. Demographic Profiles of Innovators, Noninnovators, and Others

	Innovators (*n* = *22*)	*Noninnovators* (*n* = *24*)	*Others* (*n* = *193*)
Age (mean, in years)	41.4	45.2	41.7
Mean years employed firm	12.7	16.3	12.5
Mean years in present job	2.9	3.0	3.4
Level (mean)	6.1	6.9	7.5
Gender (%)			
Female	29	4	7
Male	71	96	93
Education (%)			
Doctorate	32	21	21
Masters	41	33	26
Bachelors	23	46	48
Associate	—	—	1
High School	4	—	4
Major (%)			
Science	50	33	55
Engineering	23	17	18
Businesas	18	50	21
Other	9	—	6

Table 2. Psychological and Behavioral Comparison of Innovators and Noninnovators

	Innovators (*n* = *22*)	*Noninnovators* (*n* = *24*)	*Others* (*n* = *193*)
Psychological Indicators			
KAI Score	53.7***	43.8	47.5
Risk Score	3.5**	2.9	3.2
Self-Perception Risk	3.6**	3.0	3.5
Behavioral Indicators			
Likelihood Fund Project	4.1	3.8	3.9
Perceived Risk Level	3.1	3.3	3.1
Probability to Fund	57.2	60.1	60.2
Role Indicators			
Analyzability	2.3**	2.9	2.8
Variety	2.2**	2.9	2.7
Independent Role (%)	25	32	42
Sequential Role (%)	14*	22	20
Interdependent Role (%)	61*	46	38

Notes: $^{*}p < .05$.
$^{**}p < .01$.
$^{***}p < .001$.

The two groups differ in the expected direction in terms of the behavioral measures of risk taking, although these differences do not reach conventional levels of statistical significance. The Innovators are more inclined to fund the uncertain project (mean, 4.1 vs. 3.8, $F = 1.75$, ns), than noninnovators. Innovators perceive a lower level of risk in the situation (mean 3.1 vs. 3.3, $F = 1.05$, ns), and they require a lower probability of success before funding the project (mean 57.2 vs. 60.1, $F = .93$, ns).

The two groups differ in terms of their role perceptions. The Innovator group reports their jobs to be more uncertain. The Innovators report a lower degree of routineness (Variety mean 2.2 vs. 2.9, $F = 11.87$, $p < .001$) and a lower degree of standard procedures and practices (Analyzability mean 2.3 vs. 2.9, $F = 7.30$, $p < .01$). The two groups also report significantly different role relationships. The Innovator group reports their jobs as involving fewer sequential relationships than the Noninnovators (14% vs. 22%, $F = 4.60$, $p < .05$), and more interdependent relationships (61% vs. 46%, $F = 4.28$, $p < .05$).

DISCUSSION

This paper reported a study which has implications for the identification, placement, and development of innovators, a group of individuals believed to be essential for the continued competitive success of business organizations, especially high technology organizations. The innovators in this study were compared to noninnovators, and both groups were identified by company executives. The study then differentiated the two groups in terms of certain demographic, psychological, behavioral, and role indicators, with a view toward developing a profile of the innovators as risk takers because innovation involves risk taking. If such a profile were developed then organizations might enhance their capacity to be innovative simply by identifying and hiring risk taker/innovators.

The identification of innovators, however, is not enough because people must be effectively matched to roles in organizations. The present study did succeed in differentiating innovators from noninnovators with both psychological measures *and* role perception measures. Psychologically the innovators might be considered risk takers but, in a role sense, the innovators see themselves to be operating in somewhat uncertain environments and, as a result, may have the *opportunity* for risk taking (and the development of risk-taking attitudes) not available to their noninnovative colleagues.

The effective deployment of human resources requires the effective matching of people with roles. By focusing exclusively on innovators and their roles, however, we ignore an equally compelling match: *non*innovators and their roles. An organization does not need to be populated by too many innovators because most of what organizations do on a daily basis is routine, not innovative. Organizations need both innovators and adaptors. Even Roberts and Fusfeld (1988) define the innovative process in large organizations as one characterized by a series of interconnected roles, only some of which require "innovation" in the conventional sense. Interestingly enough, the data in the present study indicated that the innovators see their roles as being greatly interdependent with other roles. Such a view of interdependence would be necessary for Roberts and Fusfeld's view of "innovation as a series of roles" to work smoothly in a large organization.

The person/role match requirement, especially in regard to risk taking, suggests two additional considerations. First, we might consider the risk taking requirements built into roles. It may be that we don't need transcendent risk takers so much as we need people who are not averse to risk taking when the role requires it. Second, what we conventionally think of as risk taking may in fact be a capacity to operate in uncertain circumstances. Thus, we may not need to identify risk takers or gamblers so much as identify people who have the discipline (or courage) to operate in circumstances in which there are many unknowns. Or, it may not be risk takers who are as important as people who can, in Drucker's (1985) words, tolerate failure.

Finally, the results from this study suggest two things about studying risk taking. First, it is difficult to predict what people will do based on what we think they are; even based on what *they* think they are. Note that the data reported in this study indicate strong and commonly shared perceptions. That is, the group management identified as innovators (by definition a risk taking group) reported themselves to be risk takers and reported themselves capable of engaging in risk taking behaviors. Thus, to some extent the organization's assessment of "being innovative" matches the innovators' assessment of "being innovative." When given an opportunity to actually take a risk, however, these innovators did not do so. Thus, the label of "innovator" while being strongly predictive of perception and attitude is, in this study at least, less predictive of behavior.

The second issue raised by the results of this study has to do with the measurement of risk taking. MacCrimmon and Wehrung (1986) discuss at some length the likelihood that risk taking is a multidimensional construct. We agree that there are different kinds of risk, and that a person inclined toward taking physical risks (e.g., a skydiver) is not necessarily inclined toward financial risks (e.g., buying penny stocks). We tried in this study to keep the focus on job related risk taking. Nonetheless, we will not fully understand risk takers, or risk taking situations until we articulate more precisely what we mean by risk taking. It is difficult to imagine that innovators in organizations are not risk takers. But, the question remains, what kind of risks do they take? Future research will have to address more thoroughly the multidimensional character of risk taking.

APPENDIX A

Kirton Adaptation Inventory Items

1. Have an original idea
2. Proliferate ideas
3. Be stimulating
4. Stand out in disagreement against the group
5. Prefer colleagues who never rock the boat (Reversed scored)
6. Cope with several new ideas at the same time
7. Always think of something when stuck
8. Create rather than improve
9. Have fresh perspectives on old problems
10. Like the protection of precise instructions (Reversed scored)
11. Prefer changes to occur gradually (Reversed scored)
12. Often risk doing things differently
13. Prefer to work on one problem at a time (Reversed scored)

APPENDIX B

Scenario Used in the Study

As R&D manager, one of your project teams has come to you requesting an additional $100,000 in funds for a project you instituted several months ago. The project is already behind schedule and over budget, and you boss thinks the project should be dropped, but the team still believes it can be successful.

You currently have $500,000 in your budget unallocated, but which must carry you for the rest of the fiscal year. Lowering the balance by an additional $100,000 might jeopardize flexibility to respond to other opportunities.

Evaluating the situation you believe there is a fair chance the project will succeed, in which case the additional funding would be worthwhile; if unsuccessful, the money would be lost. Of the projects undertaken by this team, 30 of the last 50 have been successful.

ACKNOWLEDGMENT

This research was supported by the Center for Innovation Management Studies, Lehigh University

REFERENCES

Amabile, T. (1988). A model of creativity and innovation in organizations. In B.M. Staw & L.L. Cummings (Eds.), *Research in organizational behavior* (Vol. 10, pp. 123-167). Greenwich, CT: JAI Press.

Brockhaus, R.H. (1980). Risk taking propensities of entrepreneurs. *Academy of Management Journal, 23,* 509-520.

Drucker, P. (1985). *Innovation and entrepreneurship.* New York: Harper & Row.

Dunegan, K.J., & Duchon, D. (1989). Gender, task complexity, and risk taking: Catch 22 for women. *Mid-America Journal of Business, 4,* 23-32.

Fombrun, C., & Tichy, N. (1984). Strategic planning and human resource management: At rainbow's end. In R.B. Lamb, (Ed.), *Competitive Strategic Management* (pp. 319-332). Englewood Cliffs, NJ: Prentice-Hall.

Galbraith, J.R. (1984). Designing the innovating organization. In R.B. Lamb (Ed.), *Competitive Strategic Management* (pp. 297-318). Englewood Cliffs, NJ: Prentice-Hall.

Hannan, M.T., & Freeman, J. (1977). The population ecology of organizations. *American Journal of Sociology,* 929-964.

Hannan, M.T., & Freeman, J. (1984). Structural inertia and organizational change. *American Journal of Sociology,* 149-164.

Hisrich, R.D. (1990). Entrepreneurship/Intrapeneurship. *American Psychologist, 45,* 209-222.

Jackson, D.N. (1976). *Jackson personality inventory manual.* Goshen, NY: Research Psychologists Press.

Jackson, D.N. (1977). Reliability of the Jackson personality inventory. *Psychological Reports, 40,* 613-614.

Kanter, R.M. (1988). When a thousand flowers bloom: Structural, collective, and social conditions for innovation in organization. In B.M. Staw & L.L. Cummings (Eds.), *Research in organizational behavior* (Vol. 10, pp. 169-211). Greenwich, CT: JAI Press.

Keller, R.T., & Holland, W.E. (1978a). Individual Characteristics of innovativeness and communication in research and development organizations. *Journal of Applied Psychology, 63,* 759-762.

Keller, R.T., & Holland, W.E. (1978b). A cross validation study of the Kirton adaption innovation inventory in three research and development organizations. *Applied Psychological Measurement, 2,* 563-570.

Keller, R.T., & Holland, W.E. (1979). Toward a selection battery for research and development professional employees. *IEEE Transactions on Engineering Management, 26,* 90-93.

Kets de Vries, M.F.R. (1985). The dark side of entrepreneurship. *Harvard Business Review, 63,* 160-167.

Kirton, M. (1976). Adaptors and innovators: A description and measure. *Journal of Applied Psychology, 61,* 622-629.

Knight, K.E. (1967). A descriptive model of the intra-firm innovation process. *Journal of Business, XL,* 478-496.

Low, M.B., & MacMillan, I.C. (1988). Entrepreneurship: Past research and future challenges. *Journal of Management, 14,* 139-161.

MacCrimmon, K.R., & Wehrung, D.A. (1986). *Taking risks: The management of uncertainty.* New York: Free Press.

McKelvey, B. (1982). *Organizational systematics: Taxonomy, evolution, and classification.* Berkeley, CA: University of California Press.

Meyer, J.W., & Rowan, B. (1977). Institutional organizations: Formal structure as myth and ceremony. *American Journal of Sociology,* 340-363.

Quinn, J.B. (1985). Managing innovation: Controlled chaos. *Harvard Business Review, 63,* 73-84.

Perrow, C. (1967). A framework for the comparative analysis of organizations. *American Sociological Review, 32,* 194-208.

Peters, T., & Austin, N. (1985). *A Passion for Excellence.* New York: Random House.

Pinchot, G. (1988). Innovation through intrapreneuring. In R. Katz (Ed.), *Managing professionals in innovative organizations* (pp. 121-129). Cambridge, MA: Ballinger.

Roberts, E.B., & Fusfeld, A.R. (1988). Critical functions: Needed roles in the innovation process. In R. Katz (Ed.), *Managing professionals in innovative organizations* (pp.101-120). Cambridge, MA: Ballinger.

Romanelli, E. (1989). Organization birth and population variety: A community perspective on origins. In L.L. Cummings & B. M. Staw (Eds.), *Research in organizational behavior* (Vol. 11, pp. 211-246). Greenwich, CT: JAI Press.

Schrage, E. (1965). The R&D entrepreneur: Profile of success. *Harvard Business Review, 43,* 56-69.

Stead, D. (1990, September 23). Secrets of a cosmic cosmetician. *The New York Times,* F-25.

Steiner, G. (1988). The creative organization. In R. Katz (Ed.), *Managing professionals in innovative organizations.* Cambridge, MA: Ballinger.

Thompson, J.D. (1967). *Organizations in action.* New York: McGraw-Hill.

Waterman, R. (1988). *The renewal factor.* New York: Bantam.

Withey, M., Daft, R.L., & Cooper, W.H. (1983). Measures of Perrow's work unit technology: An empirical assessment and a new scale. *Academy of Management Journal, 26,* 45-63.

Zaltman, G., Duncan, R., & Holbek, J. (1973). *Innovations and organizations.* New York: Wiley.

DETERMINANTS OF INTENTIONS TO QUIT AMONG HIGH-TECH PROFESSIONALS

Amos Drory

The increasingly growing rate of computer applications to virtually all fields of business and commerce led to a growing job market demand for reliable Information System (IS) employees to maintain and develop information systems. Organizations found themselves intensely competing over the services of IS personnel and struggling with a chronic problem of high rate of personnel turnover. The turnover rate in this sector, which was estimated to be as high as 28% (McLauglin, 1979) caused considerable concern among researchers and practitioners alike (Albrecht, 1983; Chesebrough & Davis 1983; Kaiser, 1983).

Two studies that attempted to address the issue focused primarily on employees' job attitudes and perceptions as predictors of turnover of IS employees. Bartol (1983) found that professionalism, job satisfaction, and organizational commitment were inversely related to employee turnover on a sample of 250 computer specialists. Baroudi (1985) found that Role Ambiguity was a significant dysfunctional variable inversely related to turnover

Human Resource Strategy in High Technology.
Advances in Global High-Technology Management, Volume 1, pages 125-132.

ISBN: 1-55938-489-1

for a sample of 229 IS personnel. The purpose of the present study was to further explore the nature and causes of turnover among IS professionals in view of the unique situation in this occupational area.

It is suggested that standard models of turnover such as Steers and Mowday (1987) and Mobley (1977) may not fully account for the case of IS employees. These models assume a relatively stable situation in which employees consider alternative employment only when becoming dissatisfied with their present job. According to these models, unmet job expectations lead to negative attitudes toward the job which, in turn, trigger an active search for alternative employment.

The unique IS job market situation is characterized by high demand facing insufficient supply of qualified personnel. Companies having a growing need for information processing services engage in an aggressive competition in recruiting such employees from other firms. Consequently, IS professionals face an ongoing temptation to move from one job to another. Thus, they constantly assess the job market situation and consider alternative opportunities almost irrespective of their satisfaction with their present job.

This unique situation is hypothesized to affect turnover among IS employees in two major ways. First the motivation to stay on the job is affected to a greater extent than in more stable occupations, by job market expectations. Second, the expectations of alternative employment takes priority over their satisfaction with the present job. Thus, when the prospects for a more desirable employment are perceived as promising, a predisposition to quit will be formed primarily on this basis while giving little weight to the present job's characteristics. Only when the immediate prospects of a more promising alternative are low, resignation will be considered primarily on the basis of the quality of the present job and its characteristics.

It is therefore hypothesized that alternative employment expectations will moderate the relationships between the perceived quality of the work itself and turnover inclinations. More specifically it is suggested that job perceptions will affect turnover more strongly when the expectation to find a more attractive occupation are low than when they are high.

A second issue examined in this paper pertains to the relative impact of extrinsic as opposed to intrinsic work characteristics on turnover inclinations. It is of both theoretical and practical importance to explore the relative contribution of both variables in order to better understand the nature of turnover among IS personnel. This issue is examined on an exploratory basis without suggesting an a priori hypothesis as to the relative impact of these two categories.

METHOD

Sample

Two hundred system analysts constituted the sample for this study. The subjects were employed in several medium and large organizations representing both the private and public sector. Their mean age was 31 and their average tenure with the organization was three years.

Measures

The following paper and pencil scales were included in a questionnaire which was administered to all subjects.

- Independent Variables
  1. *Job Satisfaction.* The five subscales of the Job Descriptive Index (JDI) were used to measure job satisfaction (Smith, Kendall, & Hulin, 1969). The subscales pertain to the following aspects of work: 1 = Supervision, 2 = Co-workers, 3 = Work itself, 4 = Pay, and 5 = Promotion.
  2. *Organizational Commitment.* A 15-item Likert-scale (Mowday, Steers, & Porter, 1979). The scale is designed to measure the degree of an individual's identification with and involvement in his or her organization.
  3. *Professionalism.* A 20-item Likert-type scale developed by Kerr, Von Glinow, and Schriesheim (1977). The construct, which has been associated with turnover, consists of five dimensions: desire for professional autonomy, commitment to the profession, identification with the profession, professional ethics and collegial maintenance of standards. For the purpose of this study only the total score was used.
  4. *Turnover Expectancies.* Eight items were designed to measure the expected benefits from changing the present job. The leading question was: If you quit your present position and moved to another job how would the following aspects of your work and life be affected? The scale consisted of an intrinsic and an extrinsic subscale, each including four items as follows: intrinsic items: personal challenge, promotion opportunities, interesting job, and professional level and four extrinsic items: work conditions, supervision, job security, and income. Each item was rated on a 7-point scale: from 1 = considerable downgrading to 7 = considerable upgrading.

     The items were generated on the basis of preliminary interviews with system analysts in several organizations. In the process of developing the scale it was administered twice to a group of 43 IS employees with an interval of three weeks between administrations. The test retest correlation was .89. The Cronbach Alpha coefficient for the extrinsic subscale was .70 and for the extrinsic subscales it was .76.

- *Dependent variables*
  1. *Intention to Quit.* This attitude was measured by a three item instrument developed by Mobley (1977). All scales with the exception of Turnover Expectancies are well established in the literature with acceptable reported levels of reliability.
  2. *Procedure.* A single questionnaire containing the above instruments was given to the subjects in their workplaces accompanied with a cover letter explaining that it is administered strictly for research purposes and assuring the confidentiality of the responses. Subjects were asked not to indicate their names on the forms. Two hundred usable questionnaires were collected. In 23 cases the subjects either refused to respond or returned incomplete questionnaires.

RESULTS

Table 1 presents the intercorrelations among the study variables. It may be observed that the dependent variable (Intention to Quit) was significantly correlated with Professionalism, Commitment, Satisfaction with Work, Satisfaction with Promotion, and Intrinsic Turnover Expectancies. It may be noted that the Extrinsic Turnover Expectancies, and the Coworkers and Supervision subscales of the JDI were not correlated with the intentions to quit. The correlation with the Pay subscale of the JDI was marginally significant.

In the next step of the data analysis the combined effect of all predictors on Intention to quit was examined. A stepwise multiple regression was performed with all the attitude variables as predictors and the Intention to quit scale as the dependent variable. Table 2 presents the results of the stepwise regression. The overall multiple R obtained was .66 accounting for 43% of the variance. Four variables contributed significantly to the overall R including three job attitudes variables and the measure of Intrinsic Turnover Expectancies. Satisfaction with promotion was also marginally significant.

The second hypothesis concerning the moderating effect of turnover expectancies on the relationship between job attitudes and intention to quit was tested by means of the moderated regression approach (Zedeck, 1971). Following this procedure a multiple regression is conducted on the dependent variable (Intention to Quit) with the independent variable followed by the cross product of the independent variable and the moderator as predictors. The contribution of the cross product variable to prediction reflects the moderating effect. The moderating variable according to the second hypothesis was

Turnover Expectancies. The moderating effect was tested with regard to three job attitude variables: (1) Professionalism, (2) Organizational Commitment, and (3) Satisfaction with Work. Table 3 summarizes the moderating effects. The results suggest that the hypothesized moderating variable yielded a significant effect in all cases. An examination of the plots of the relationships between each of the three independent variables and the dependent variable when moderated by the moderating variable suggested that the differences were in the hypothesized direction.

Table 1. Pearson Correlation Among the Study Variables*

	2	*3*	*4*	*5*	*6*	*7*	*8*	*9*	*10*
1. Intention to Quit	−.31	−.59	−.41	−.32	−.12	.04	−.08	.39	.10
2. Professionalism		.48	.41	.41	.41	.29	.07	−.24	−.14
3. Commitment			.39	.33	.04	.10	.11	−.40	−.21
4. JDI Work				.36	.04	.31	.16	−.33	−.09
5. JDI Promotion					.03	.04	.16	−.22	−.05
6. JDI Pay						.05	.04	−.13	−.12
7. JDI Coworkers							.36	−.07	−.16
8. JDI Supervision								−.14	−.11
9. Intrinsic Turnover Expectancies									.33
10. Extrinsic Turnover Expectancies									

Note: *Coefficients of above .12 are significant at < .05; coefficients of above .16 are significant at <.01; and coefficients of above .22 are signicant at <.001.

Table 2. Stepwise Multiple Regression With All Independent Variables on Intention to Quit

Variable	*R*	R^2	*Significance of Predictor*	*Overall Significance*
Commitment	.59	.35	.000	.000
Professionalism	.61	.38	.000	.000
JDI Work	.63	.40	.000	.000
Intrinic Turnover Expectancies	.65	.42	.020	.000
JDI Promotion	.65	.42	.069	.000
JDI Supervision	.65	.42	NS	.000
JDI Pay	.65	.42	NS	.000
JDI Coworkers	.66	.43	NS	.000

Table 3. Regression of Moderating Variables on Intention to Quit

	R^2 of Variable	*Significance of Increment*	*Overall R*
Professionalism (independent variable)	.10	.000	.10
Professionalism × Intrinsic Turnover Expectancies (moderator)	.11	.000	.21
Work Satisfaction (independent variable)	.17	.000	.17
Work Satisfaction × Intrinsic Turnover Expectancies (moderator)	.06	.000	.23
Commitment (independent variable)	.35	.000	.35
Commitment × Intrinsic Turnover Expectancies (moderator)	.03	.002	.38

DISCUSSION

The results of this study support previous findings concerning the relationships between the intention to leave the organization and job attitudes (Baroudi, 1985; Bartol, 1983). Further, the present findings suggest that the distinction between intrinsic and extrinsic job characteristics as originally suggested by Herzberg (1966) is of particular importance for IS employees. Turnover intentions were associated predominantly with intrinsic job characteristics and with intrinsic turnover expectancies. Forty-three percent of the variance of the attitude toward quitting was accounted in this study by intrinsic work variables.

The hypothesis concerning the moderating effect of turnover expectancies was fully supported by the present findings. The findings suggest that IS employees tend to give priority to the assessment of turnover prospects and benefits in making decisions about job change. When the prospects for alternative employment are perceived as good, the assessment of the present job characteristics is given less weight in forming an attitude about quitting. As the perceived prospects for alternative employment get lower, the employee gives more weight to the characteristics of the present job. Thus a better understanding of turnover among IS employees may be gained by considering a moderated rather than an additive model of the decision-making process which leads to turnover.

Some practical implications concerning the retention and attraction of IS employees may be suggested in view of the present findings. The relative importance of intrinsic work characteristics for turnover decisions implies that a better retention of IS employees may be obtained through a job design approach. Although information systems tasks are often rich and meaningful this is not always the case. Job design is too often determined intuitively without any particular concern for its motivational implications. The extent to which individual jobs contain enrichment elements of autonomy, feedback, task identity, skill variety, and task significance varies in this occupation category among jobs and organizations. The implementation of job enrichment models such as Hackman and Oldham's (1976) may enhance both work satisfaction and a sense of professionalism among IS employees. In addition, further efforts should be devoted to the development of specific enrichment models for IS jobs.

Organizational commitment may be enhanced through employee participation and involvement in meaningful organizational decision making beyond the technical scope of the job. The relationships between promotion satisfaction and turnover intentions suggests that long-term career planning for IS personnel may also contribute to better employee retention among this occupational group.

The meaningful impact of perceived job opportunities on the predisposition to leave cannot be directly controlled by the organization. However, a greater emphasis on careful job enrichment and on future career opportunities may help to counter-balance the influence of perceived alternatives. It is believed that such focus on job design and career development may also improve the attractiveness of the firm in the eyes of new applicants.

Finally, two limitations of the present study are worth noting. First, the dependent variable in this study was the desire to quit rather then actual turnover. As such, it only serves as an important but indirect measure of turnover. Further research is needed to relate the present propositions to actual turnover data.

Second, this study focused on IS employees. In order to further validate the present findings, future research should examine the same propositions on other 'hot' occupational groups and compare them to standard occupations. Such expansion may further contribute to our understanding of the decision process leading to turnover behavior.

REFERENCES

Albrecht, M.J. (1983). Managing employee careers. *Computerworld, 17,* 21-32.

Baroudi, J.J. (1985). The impact of role variables on IS personnel work attitudes and intentions. *MIS Quarterly, 9,* 341-356.

Bartol, K.M. (1983). Turnover among DP personnel: A causal analysis. *Communications of the ACM, 26,* 807-811.

Chesebrough, P.H., & Davis, G.B. (1983). Planning a career path in information systems. *Journal of Systems Management, 34,* 6-13.

Hackman,J.R., & Oldham, G.R. (1976). Motivation through the design of work: Test of a theory. *Organizational Behavior and Human Performance, 16,* 250-279.

Herzberg, F. (1966). *Work and the nature of man.* Cleveland, OH: World.

Kaiser, K.M. (1983). DP career path. *Datamation, 29,* 178-185.

Kerr, S., Von Glinow, M.A., & Schriesheim, J. (1977). Issues in the study of Professionals in organizations: The case of scientists and engineers. *Organizational Behavior and Human Performance, 18,* 329-345.

McLauglin, R.A. (1979). That old bugaboo turnover. *Datamation, 25,* 97-101.

Mobley, W.H. (1977). Intermediate linkages in the relationship between job satisfaction and employee turnover. *Journal of Applied Psychology, 62,* 237-240.

Mowday, R.T., Steers, R.M., & Porter, L.W. (1979). The measurement of organizational commitment. *Journal of Vocational Behavior, 14,* 224-247.

Smith, P.C., Kendall, L.M., & Hulin, C.L. (1969). *The measurement of satisfaction in work and retirement: A strategy for the study of attitudes.* Chicago: Rand-McNally.

Steers, R.M., & Mowday, R.T. (1987). Employee turnover in organizations. In R.M. Steers & L.W. Porter (Eds.), *Motivation and work behavior* (pp. 394-406). New York: McGraw-Hill.

Zedeck, S. (1971). Problems with the use of "moderator" variables. *Psychological Bulletin, 76*(1), 295-310.

MANAGING THE DUAL LADDER: A LONGITUDINAL STUDY

Ralph Katz, Michael L. Tushman, and Thomas J. Allen

Organizations employing many professional specialists, engineers and scientists in particular, face the dilemma of establishing a reward system that is both stimulating to the professional and productive for the organization. This problem stems, in part, from the notion that specialist groups bring to the organization a set of attitudes and career aspirations that are in conflict with the organization's work requirements and career paths. As argued by Kornhauser (1962) and Hall (1968), many R&D professionals are socialized into their technical occupations with values and definitions of success that differ significantly from those prevailing in the traditional managerial setting. In the typical organization, for example, management expects authority to be discharged according to the hierarchical principle, delegated through a series of well-ordered job positions. Technical professionals, however, come to value the freedom to pursue their own technical interests, the responsibility for making judgments in areas of technical competence, and the ability to exercise control through their peer group.

Human Resource Strategy in High Technology.
Advances in Global High-Technology Management, Volume 1, pages 133-150.

ISBN: 1-55938-489-1

For many years now, much has also been written about how professional incentives clash with those organizational incentives normally available to managers (for recent reviews, see Kaufman, 1982; Raelin, 1985; Von Glinow, 1988). In theory, many technical professionals are supposedly motivated by a desire to contribute to their fields of knowledge and to establish distinguished reputations within their technical disciplines. In a sense, they are strongly oriented toward work in their professions, developing strong commitments to their specialized skills and outside professional reference groups (Gouldner, 1957). Such a "cosmopolitan" orientation, however, often leads to less organizational loyalty. Managers, on the other hand, desire upward mobility in the organizational hierarchy. In a sense, they are more committed to developing their own "local" organizational careers. They do this by focusing more on the achievement of company objectives and the acquisition of organizational approval and promotion. As one R&D professional recently framed it for us: "To have my ability recognized rather than my authority is far more rewarding." To the true professional, then, upward mobility in the managerial hierarchy is of little importance compared to autonomy in the practice of one's technical specialty. Success is, therefore, defined independent from managerial advancement. In short, the argument is that professionals acquire status and define success from the perspectives of their technical colleagues while managers build these same attributes from the perspectives of their organizational superiors.

Whether engineers and scientists in technology-based organizations are really more interested in peer recognition than they are in organizational advancement has been the subject of much debate. Many studies, including Allen (1977) and Ritti (1971), have shown, for example, that engineers are very different in their professional and organizational orientations from their more scientific counterparts. The findings of Kerr and Von Glinow (1977), Bailyn (1980), Schein (1988), and others also question whether one can truly generalize within any professional occupation while the results of Dalton, Thompson, and Price (1982) and Allen and Katz (1986) indicate that orientations can change significantly over time or through different career stages. Nevertheless, there is general agreement that within most organizations, there is some proportion of professionals who prefer technical problem-solving and for whom management has very little attraction.

THE DUAL LADDER STRUCTURE

Despite these purported "professional/managerial" differences, the highest rewards in most business organizations are conferred on those who assume additional managerial responsibility. Advancement up the managerial ladder secures increases in status, recognition, salary, influence, and power. For many professionally-oriented technologists, movement into management becomes the most viable career strategy simply because their opportunities to achieve success without undertaking such managerial responsibilities are very limited. As a result, many productive engineers and scientists feel frustrated as they are "pressured" to take on managerial and administrative roles they really do not want in order to attain higher salary and more prestige.

In addition to this source of individual frustration, the organization itself may also be adversely affected by its failure to provide alternative rewards for those technical professionals who either do not aspire to or who show no aptitude for management. As argued by Steiner (1988), Allen, Katz, Grady, and Slavin (1988), and many others, long-term technology-based innovation requires the continued productive efforts of experienced engineers and scientists who will remain creative, current, and productive in their disciplines. The reluctant movement of highly competent individual contributors into management can seriously deplete the organization's pool of creative technical talent. Not all high performing technologists, moreover, have the interpersonal, communication, or leadership skills necessary to make an effective transition from technical specialist to technical manager (Badawy, 1988). Nevertheless, if technical professionals see managerial incentives as the only real path to higher pay and higher status, then these professionals will also feel compelled to become managers. The organization, as a result, not only loses their technical abilities but risks additional problems and discontent as their success as managers diminishes.

The "dual ladder system of career advancement" is an organizational arrangement that was developed to solve these individual and organizational problems by providing meaningful rewards and alternative career paths for organizational professionals (Kaufman, 1975). Generally speaking, the dual ladder approach is the formalization of promotions along two parallel hierarchies. One hierarchy provides a managerial career path while the other provides advancement as a professional or technical staff member. They promise equal status and rewards at equivalent levels in the two hierarchies. Also known as the "technical" or "individual contributor" ladder, the dual ladder was established over forty years ago to reward professionals (especially scientists and engineers) for outstanding scientific and technical performance without having to remove them from their professional work (Shepard, 1958). By providing professionally-oriented individuals with an opportunity and incentive to remain active in their fields and to stay up-to-date, the dual ladder aims to secure for the high technology firm an adequate pool of technical talent.

Although dual ladders have now been in use for some time, their success has been the focus of much agonizing debate. A critical tone pervades the literature, and a whole host of problems have been identified (Allen & Katz, 1989; Roth, 1988). Most cultures, for example, automatically associate prestige with managerial advancement. Titles of Department Head and Vice President convey images of success while titles of Senior Researcher and Lead Engineer are considerably more ambiguous and therefore more subject to skepticism.[1] Many organizations exacerbate these differences by not living up to their promised commitments of creating equal status, perquisites, resources, and other financial and symbolic rewards to those of equivalent levels in the managerial and professional hierarchies. Frequently too, management does a poor job of publicizing the technical ladder and little observable change takes place either in work activities or responsibilities after technical promotion. An additional problem arises when technical promotions are debated through justifications of past contributions while managerial promotions are more positively discussed in terms of future promise and potential.

Another set of problems concerns the nature of incentives associated with each ladder. Movement up the managerial ladder usually leads to positions of increased influence and power within the organization. The number of employees under a manager typically increases with promotion and such resources can be mobilized more easily to carry out the manager's needs and demands. In sharp contrast, advancement up the technical ladder usually leads to increased autonomy in the pursuit of one's technical interests but often at the expense of organizational influence and power (Allen & Katz, 1989; Goldner & Ritti, 1967). Neither the number of subordinates nor any visible means of power increase, fostering perceptions that the technical ladder might really be less important. The issue of relevance becomes even more difficult as the organization grants professionals enough freedom to select their work with little linkage between their activities and company objectives, returns, or paybacks. Such conflicts are aggravated even more as the organization chooses to either eliminate or deemphasize certain areas of interest. As a result, supervision of individual contributors becomes more difficult and feelings of isolation from the organization become more pronounced. According to Emmons (1977), the risk is that the technical side becomes a "parking lot" for bright technologists whose abilities to generate ideas easily outstrips the capability of the organization for dealing with them. The rewards of freedom and independence can also bring with them feelings of rejection and disconnection.

Finally, there is the inevitable tendency to "pollute" the technical side of the dual ladder. In addition to rewarding outstanding technical performers who choose to remain in the organization as individual contributors, the technical ladder becomes a repository for less successful, unnecessary, and even incompetent managers. Over time, the criteria for technical promotion are gradually corrupted to encompass not only technical contributions but also organizational loyalty, rewarding those individuals who have been "passed over" for managerial positions. Another common practice is to use the professional ladder primarily for pacifying individuals who are technically competent and who deserve to be rewarded, but who lack diplomatic skills or management ability. When any of this is done, it can make the technical ladder into a consolation prize, demotivating individuals who interpret technical promotions not as a reward but simply as a signal that they are "not good enough to be a manager." Certainly, such misuses undermine the integrity of the dual ladder system.

Much has been done over the past few years to improve the formal structures of dual ladder systems to alleviate these problems. Using internal and external peer reviews, organizations have begun "policing" their technical ladders to protect their purity and prevent the "dumping ground" abuses. They have tried to strengthen their commitment to the technical side through increased publicity, recognition, career counseling, and information dissemination; through making the ladders more comparable in numbers of people, and perquisites at equivalent hierarchical levels; through clearer job descriptions, qualifications, responsibilities, performance standards, and reporting relationships; and through greater involvement in organizational decision-making and in influencing technical strategy.

EXPLORING THE PROMOTIONAL DYNAMICS OF DUAL LADDERS

Despite these design changes, we still know very little about how technical professionals make sense out of the dual ladder system. One of the most persistent themes in organizational theory is that individuals build perceptions through their interactions and experiences with others (Salancik & Pfeffer, 1977; Weick, 1969). According to this social constructionist point of view, the success or failure of a dual ladder system will be defined not so much by structural features per se but by the actual promotional dynamics occurring within the organization. If the "better" people move along a particular ladder, for example, then perceptions of success or the "right" track are more likely to be defined by this pattern.

Underlying this argument is the idea that scientists and engineers do not have well-defined notions of success and failure as they graduate and enter organizations to practice their technical specialties. Colleges and universities do not provide them with good definitions. Instead, success is characterized in very general terms such as "advancement," "getting ahead," or "making a difference." Better definitions of success, therefore, must emerge from the organizational context in which individuals find themselves. Without a firm prior definition, how technologists come to view the organization's technical and managerial ladders will be influenced more strongly by their organizational contacts than by their professional socialization during their years of study.

According to the results of many studies, the most important set of contacts affecting one's organizational career success develops through the mentoring relationship of one's immediate supervisor (Kanter, 1977; Kram, 1985). As discussed by Schein (1978) and Katz (1988), a professional's boss plays a critical role in helping to develop a more accurate assessment of an individual's skills and abilities, in understanding the norms and values of the local organizational context, and most important, in defining how the individual becomes socialized and accepted within the setting as a contributing member. Studies have also suggested that working for supervisors who are themselves promoted enhances the promotion chances of their subordinates (Katz & Tushman, 1983; Webber, 1976).

If supervisors influence the career outcomes of their technical subordinates, then engineers and scientists working for supervisors promoted along different ladders might, themselves, be affected differently in their subsequent promotions. Technical professionals working for supervisors who receive a managerial promotion, for example, might then follow a very different pattern of promotions from their technical peers working for supervisors who receive a technical ladder promotion.

This research will investigate the career outcomes of technical professionals reporting to supervisors who were promoted on the two ladders. The key issue is not only whether one's chances for promotion are improved by working for more influential and promotable bosses, but whether and how promotional paths are affected.

The Influence of Gatekeeping Supervisors

Research in RD&E environments consistently shows that technological gatekeepers are extremely important in the effective transfer and utilization of external technology and information (Allen, 1977). Allen and Cohen (1969) define gatekeepers as those technical professionals who are strongly networked to both internal colleagues and external sources of technical information. They are also high technical performers who are interpersonally approachable and helpful. Findings by Tushman and Katz (1980) indicate that such individuals have a significant effect on the overall technical performance of development projects, while Katz and Tushman (1983) show their strong impact on the long-range retention of young technical professionals. Since many gatekeepers are also project supervisors, this study will also contrast how the promotional patterns of gatekeeping supervisors, in comparison to the other supervisors, influence the career outcomes of their technical subordinates.

The Influence of Communication

Over the past 20 years, empirical studies have demonstrated that interpersonal communication, rather than the written media, is the principal means by which technical professionals collect and disseminate important ideas and information. As shown by Graen and Ginsburgh (1977) and Katz and Tushman (1983), it may not be the supervisor's hierarchical status per se that is important in enhancing the career outcomes of subordinates. More beneficial are the higher levels of work-related communication that are fostered by certain supervisors. Katz and Tushman (1983), for example, showed just how important gatekeeping supervisors could be in facilitating higher levels of internal communication to reduce turnover. In a related study, Katz and Tushman (1981) found that promotions of project supervisors were strongly related to prior communication patterns. A higher level of internal communication was positively associated with managerial promotions but more isolated supervisors were promoted on the technical side. In a similar vein, this study will also examine the communication behaviors of technical professionals to see if supervisory relationships systematically affect either internal or external patterns of interaction, and whether any observed differences can be linked to promotional differences within the dual ladder system.

The Influence of Project Task Characteristics

One of the conclusions produced by research on the management of technology is that not all RD&E projects are alike in the way they function on in the way they should be managed. Because of strong differences in work requirements, there are substantial communication and information processing differences among groups involved in research, development, and technical support activities (Tushman, 1979). External communication with outside R&D colleagues, for example, has been found most important in applied research, while contact with other corporate areas (e.g., marketing, manufacturing, and finance) is most important in development (Allen, Tushman, & Lee, 1979). Technological gatekeepers, moreover, are most important in development work but not particularly helpful in research areas (Tushman & Katz, 1980). As a result of such important task differences, the influence of supervisory promotions and communication patterns on the career outcomes of technical professionals will be systematically examined in the research, development, and technical support areas.

Technical Performance

Technical creativity and performance in the past and present are the usual criteria for promotion to the technical ladder. Organizations hope, however, that these individuals will continue to be high technical performers and will continue to contribute technically to the organization. Many companies, however, are disappointed with the continued technical contributions of professionals promoted on the technical ladder. As a result, this study will also try to ascertain whether particular work experiences are especially effective in keeping technologists productive in a dual ladder system. Are technologists promoted from research settings to the technical side more likely to remain high performing than their counterparts from development and technical support areas? Given the problem of technical obsolescence, does age have any effect on a person's ability to remain productive? And finally, just how important is one's network of contacts, both inside and outside the company, for remaining a strong individual contributor in an organization whose work assignments are highly interdependent?

RESEARCH METHOD

This study was conducted among the 345 technical staff members of a large RD&E facility, divided into seven separate functional departments. A total of 61 projects, organized around different disciplines and product-based problems were identified across these departments. Only professionals assigned to these projects at the beginning of our data collection participated in this study. Each technologist was a member of only one project and worked for a single project supervisor.

Communications and Gatekeeping Supervisors

As described by Katz and Tushman (1979), each professional reported (on specially provided lists) those individuals with whom they had work-related communications on a randomly chosen day each week for 15 weeks. Using this method, we could accurately obtain for each technical professional six mutually-exclusive measures of communications as follows:

1. *Departmental communication:* The amount of communication with technical peers within the individual's functional department (including project).
2. *Laboratory communication:* The amount of communication with technical colleagues within the other six functional departments.
3. *Supervisory communication:* The amount of communication with the individual's immediate project supervisor.
4. *Managerial communication:* The amount of communication with the department manager.
5. *Communication with other functions:* The amount of communication with individuals outside the RD&E facility but within other corporate divisions, primarily marketing, manufacturing, and finance.
6. *External communication:* The amount of communication with outside RD&E professionals.

Since gatekeepers are defined as having high internal and external networks, this study defined gatekeepers empirically as those project members for whom both departmental and external communications were both in the top fifth of their respective distributions. This definition is consistent with prior studies, including, Allen (1977), Whitley and Frost, (1973), and Tushman and Katz (1980). Based on these calculations, 15 (25%) of the project supervisors were also functioning as technological gatekeepers while 46 (75%) of the project supervisors were not.

Project Characteristics

Using definitions described in Katz and Tushman (1979), respondents indicated how well the objectives of their work fell into the categories of research, development, or technical support. As in Pelz and Andrews (1966), project members also indicated the percentages of their project activities that fell into each of these three possible project categories. A weighted average of these two answers is used to calculate a score for each project member. To categorize projects reliably, these individual scores are then averaged within projects to yield a classification of 14 research, 24 development, and 23 technical support projects.

Promotion and Performance

Five years after the collection of the data on communication and reporting relationships, the authors returned to the organization to collect promotional data for all the original participants (i.e., all technical project members and project supervisors). In this organization, all technical and managerial ladder positions and titles start above the project supervisory level. Because the nature of our study had not involved anonymity, we could easily determine from the facility's personnel list of names and titles the promotional histories of our participants over the five years. For each technical professional and project supervisor, we could determine whether they were: (1) promoted on the technical ladder, (2) promoted on the managerial ladder, (3) not promoted above the project level, or (4) had left the organization. Finally, several high-level managers, who were concerned about the overall effectiveness of dual ladders were asked to evaluate the current technical contributions of those who had been promoted along the technical side and for whom they had sufficient knowledge. A 4-point Likert scale ranging from low to high was used with the ratings given by several managers averaged to yield overall performance scores. Unfortunately, comparable performance ratings for those promoted along the managerial ladder could not be obtained.

RESULTS

In this type of longitudinal study, it is important to analyze samples of individuals who are comparable. In the ideal case, one should conduct a cohort-type of analysis, comparing professionals who are as similar in age and organizational experience as possible. In this organization, the ages of technical professionals promoted on both the technical and managerial ladders covered too wide a spectrum, ranging from the mid-20s to the mid-50s. Further examination of these distributions, however, reveals that 70% of those promoted on the technical side were in the 25 to 45 age range at the start of our study while 70% of those promoted managerially were in the 27 to 35 age range. All of the analyses reported here, therefore, are based on samples of professionals who fell within these age ranges at the start of our study and who remained within the organization over the succeeding five years. Despite the organization's strong growth during the 5-year interval, almost 37% of the original respondents had left, an average of about 79 per year.

Previous studies have hinted that gatekeeping supervisors may play a strong role in the personal growth and development of project members, providing them with better work opportunities, increased exposure, and more effective information networks. The results (Table 1) show how one's career outcomes are related in this organization to working for a gatekeeping supervisor. As reported in Table 1A, 3.5% of those professionals working for a gatekeeping supervisor at the start of our study were later promoted on the technical ladder. Surprisingly enough, over four times as many (14.1%) of those not working for a gatekeeping supervisor received technical promotions. In fact, of the 18 professionals who received a technical ladder promotion, only one had been working for a gatekeeping supervisor.

Table 1. Proportions of Engineers Promoted on Technical and Managerial Ladders as a Function of Their Supervisors' Gatekeeping Role

Prior Role of Supervisor	*Proportion Promoted* (%)	*N*	*P*
A. Engineers in the 27 to 45 Age Range in which 70% of the Technical Promotions Occurred			
Assigned to Gatekeeping Supervisor	3.5	28	
			0.05
Not Assigned to Gatekeeping Supervisor	14.1	121	
B. Engineers in the 27 to 35 Age Range in which 70% of the Managerial Promotions Occurred			
Assigned to Gatekeeping Supervisor	41.2	17	
			0.05
Not Assigned to Gatekeeping Supervisor	15.2	46	

A very different pattern appears for management promotions (Table 1B). Looking at those professionals who had worked for a gatekeeping supervisor, over the following five years one discovers that 41.2% received a management promotion, while only 15.2% of those not assigned to a gatekeeping supervisor received such a promotion—a significant difference in a direction opposite to that of Table 1A. In this organization, half of the 14 engineers receiving managerial promotions, in the five year period, had reported to a gatekeeping supervisor. Since all but one of the gatekeeping supervisors in the organization were promoted on the management ladder, it is also important to see how the career outcomes of professionals working for gatekeepers compares with other possibilities. This includes working for supervisors who were (1) promoted managerially, (2) promoted technically, (3) not promoted, or (4) had left the organization. In contrast to the findings for gatekeeping supervisors, the data (Table 2A) show that a large percentage (41.7%) of those professionals working for supervisors promoted on the technical ladder were themselves promoted technically. This is more than four times the rate for the other three categories, each of which is below 10%. In this organization, over half of those professionals promoted technically were working for a supervisor, who not only was not a gatekeeper, but who in addition was on the technical promotion

Table 2. Proportions of Engineers Promoted on Technical and Managerial Ladders as a Function of Their Supervisors' Promotion Record

Supervisors' Career Progress	*Proportion of Subordinates Promoted (%)*	*N*	*P*
A. Engineers in the 27 to 45 Age Range in which 70% of the Technical Promotions Occurred			
Promoted Managerially	7.7	65	0.01
Promoted Technically	41.7	24	0.01
Not Promoted	9.1	11	0.01
Left the Organization	5.6	36	0.01
B. Engineers in the 27 to 35 Age Range in which 70% of the Managerial Promotions Occurred			
Promoted Managerially	30.0	30	N.S.
Promoted Technically	9.1	11	N.S.
Not Promoted	11.1	9	N.S.
Left the Organization	23.1	13	N.S.

track. The results from Table 2B, on the other hand, are not nearly as strong as those found in Table 1B. Only 30% of the engineers reporting to supervisors promoted managerially received a similar promotion. This reduced rate stems from the fact that only about 15% of those working for nongatekeeping supervisors promoted managerially were also promoted to management positions. This low rate is comparable to the rates of engineers reporting to supervisors promoted technically or not promoted. It is even less than the rate of those whose supervisors had left. Gatekeeping supervision is by far the most instrumental factor for high rates of managerial promotion, fostering about 78% of the nine promotions received by individuals who reported to supervisors promoted managerially.

The data in Table 3 are the same results as those in Table 2A but analyzed by the project areas in which professionals had worked. Although the sample sizes are small, the highest rate of technical promotion occurs for those who worked on applied research projects for supervisors promoted on the technical ladder. Two-thirds of these professionals received technical promotions over the 5-year period, a rate considerably higher than engineers in the other 11 cells. Contrastingly, with respect to managerial promotions, those professionals who had worked on development projects for gatekeeping supervisors received the highest rate of managerial promotion (Table 4). Two-thirds of them received management positions. One must be careful, however, in interpreting this result, since sample sizes are again very small. Nevertheless, this rate is

Table 3. Proportion of Engineers Promoted on the Technical Ladder as a Function of Their Supervisors' Promotion Record and Previous Area of Work

	Supervisor's Career Progress			
Area of Work	*% Promoted Managerially*	*% Promoted Technically*	*% Not Promoted*	*% Left the Organization*
Applied Research	13.3*	66.7*		0.0*
	($N = 15$)	($N = 12$)	($N = 0$)	($N = 7$)
Product or Process	5.0	12.5	0	6.7
Development	($N = 40$)	($N = 8$)	($N = 5$)	($N = 15$)
Technical	10.0	25.0	16.7	7.1
Support	($N = 10$)	($N = 4$)	($N = 6$)	($N = 14$)

Notes: Analysis based on engineers in the 27 to 45 age range in which 70% of the technical promotions occurred.
*$p < 0.01$.

Table 4. Proportion of Engineers Promoted into Management as a Function of Their Supervisors' Gatekeeping Role and Their Previous Area of Work

	Type of Supervisor		
Prior Area of Work	*Gatekeeper*	*Other*	*P*
Applied Research	33.3 ($N = 6$)	30.0 ($N = 10$)	N.S.
Product or Process Development	66.7 ($N = 6$)	14.8 ($N = 27$)	0.05
Technical Support	20.0 ($N = 5$)	10.0 ($N = 10$)	N.S.

significantly higher than the managerial promotion rates for any other project or supervisory category (the rate in development work with a nongatekeeping supervisor, for example, was only 14.8%).

Clearly, gatekeeping supervisors have significant impact on the managerial promotion rates of their technical subordinates, especially in product or process development. What is it that brings about these higher rates of promotion? As previously discussed, gatekeepers may influence the communication networks of other professionals, enhancing their exposure and linkages to important information sources. An analysis is, therefore, made to ascertain whether engineers reporting to gatekeepers have significantly different interaction patterns than those not reporting to gatekeeping supervisors. The results portray a strong influence by gatekeepers over the communication activity of their technical staff except in the applied research area (Table 5). Engineers reporting to gatekeepers have much more contact with their departmental colleagues and their gatekeeping supervisors than those engineers working for a non-gatekeeping supervisor. More important, however, is the finding that in both the development and technical support areas, engineers working for gatekeepers have four times as much contact with their department managers. It appears that gatekeeping supervisors can strongly promote the hierarchical integration of their technical staffs.

Finally, we wished to see if we could get some better insight into why some engineers remain more effective technical contributors on the technical ladder. An examination of the current technical performance data for professionals promoted to the technical track did not reveal any significant differences in current performance either by age or by nature of work. In addition, none of the prior measures of communication relate significantly to current technical performance except for communications with other functions. There is a strong positive relationship between the degree to which professionals interacted with other corporate areas and the extent to which they were contributing effectively to the organization five years later (Table 6). This relationship remains strong even when reexamined by age and by project area.

Table 5. Mean Level of Engineers' Communication as a Function of Supervisor's Gatekeeping Role and Nature of Work

	Communications per Month		
Direction of Communication	*Applied Research*	*Product/Process Development*	*Technical Suport*
Departmental Colleagues			
Engineers with a Gatekeeping Supervisor	99.3	193.0*	153.7*
Engineers without a Gatekeeping Supervisor	93.6	109.7*	91.8*
Laboratory Colleagues			
Engineers with a Gatekeeping Supervisor	28.8	23.5	11.2
Engineers without a Gatekeeping Supervisor	22.7	22.6	10.4
Immediate Supervisors			
Engineers with a Gatekeeping Supervisor	24.3	26.5*	28.7*
Engineers without a Gatekeeping Supervisor	21.8	17.3*	19.2*
Department Manager			
Engineers with a Gatekeeping Supervisor	5.6	8.5**	10.9**
Engineers without a Gatekeeping Supervisor	3.4	2.1**	2.7**
Other Functions in the Company			
Engineers with a Gatekeeping Supervisor	3.8	24.9	19.6
Engineers without a Gatekeeping Supervisor	5.8	21.6	24.0
External Professionals			
Engineers with a Gatekeeping Supervisor	3.2	1.8	0.6
Engineers without a Gatekeeping Supervisor	2.6	1.1	1.0

Notes: Analysis based on engineers in the 27 to 35 age range in which 70 percent of the managerial promotions occurred.
*$p < 0.05$.
**$p < 0.01$.

Table 6. Crosslag Correlation Between Communication with Other Functional Areas in the Company and Performance Five Years Later for Engineers Promoted on the Technical Ladder

	N	*r*	*P*
All Engineers Promoted on the Technical Ladder	18	0.67	0.01
From Applied Research	9	0.81	0.01
From Development and Technical Support	9	0.64	0.05
Engineers			
Under 39 Years	9	0.74	0.01
39 or Older	9	0.65	0.01

DISCUSSION

The research findings reported here demonstrate very clearly just how much the relationships and contacts taking place within a given organizational context can affect the dynamics of a dual ladder system. As hypothesized, professionals do not have well-defined preconceived notions of success upon graduation; instead, their careers and perceptions develop from their organizational experiences. In particular, our results suggest that first-level supervisors play a strong role in the shaping of the professional's outlook toward the dual ladder system. In this organization, there were very different patterns of influence on the promotions of individuals to the technical and managerial ladders. Those supervisors who were promoted on the technical ladder strongly affected their subordinates who tended to follow that path, especially if they were working in applied research. Very few managerial promotions took place for professionals reporting to these supervisors; in fact, only one such promotion occurred over the five years. In contrast, professionals working for gatekeeping supervisors were not only more likely to be promoted but were more likely to assume management positions. Very few of the professionals reporting to gatekeeping supervisors received a promotion on the technical side. There was, as before, only one such occurrence.

By examining these kinds of dynamics, organizations can gain insight into the success of their dual ladder systems. The promotional data reported in this paper not only support the disproportionate influence that supervisors have on the career outcomes of their subordinates, but also point out the concerns that can emerge in a dual ladder system if these influence patterns become highly restrictive. Technical professionals in this organization, for example, who value or want a managerial career but who find themselves working for supervisors "going technical" may feel frustrated and disenchanted as they see less visible support for their career choice. Similar dissatisfaction could emerge for those professionals wishing to remain individual contributors but who find themselves working for supervisors who strongly favor the managerial side. In a dual ladder system, such mismatches need to take place if organizations are to fully utilize the talents of their technical work force across the full spectrum of RD&E activity. Past research also shows that supervisors promoted technically are less well integrated into the communication network within their organizations than their managerially promoted counterparts. If these supervisors are also strongly instrumental in the socialization of those professionals who follow them to the technical side, then it is likely that such isolation will become self-sustaining. Perhaps this is a reason why so many companies feel that those on the technical ladder are disconnected from what is going on in the organization. If the technical ladder involves greater freedom over what one does, and if those on the ladder have less interaction with others in the organization, the tendency may be to choose work that is even less interdependent (i.e., less mainstream) thereby aggravating the situation even more.

Much of the controversy surrounding dual ladders revolves around the issue of power: those on the management side have it; those on the technical side do not. The promotional patterns revealed in this study highlight part of this problem. Given the important organizational and technical roles that gatekeepers play in their environments, they represent a strong source of both formal and informal power. It makes sense, therefore, that the ladders these individuals choose to follow will greatly affect the allocation of power. In this organization, all but one of the technological gatekeepers were promoted to the management track. Such a one-sided situation is probably not conducive to a meaningful and equitable distribution of power across the two ladders. How gatekeepers are promoted and distributed across the dual ladder may be very critical to the ultimate success of the organization's dual ladder system.

Finally, our findings suggest that professionals promoted to the technical side are more likely to remain effective if their work experiences prior to promotion are not narrowly defined but involve interaction with other corporate functions. This was a very robust finding in that the relationship held up in all sub-analyses. There may be many alternative explanations for this strong relationship. By being exposed to other parts of the organization, perhaps these individual contributors were able to develop linkages to those places that became important sponsors or recipients of their work. Perhaps these experiences gave them a greater awareness of how to present ideas and information more effectively to other parts of the organization. Although many such possibilities exist, the beneficial effects of these kinds of cross-functional interactions need to be validated and better understood, for the ultimate success of a formal technical ladder will depend on the organization's perceptions of the "returns" from this population. Finally, readers must note that all of the analyses presented in this paper may be idiosyncratic to this organization. Nevertheless, it is through studies of the present sort that we will learn more about the true dynamics of dual ladders so organizations can take more specific action to improve their systems and the phrase "the dreaded dual ladder" will not be so commonly heard.

ACKNOWLEDGMENT

The research on which this paper is based was supported, in part, by a grant from the U.S. National Science Foundation through the Lehigh University Center for Innovation Management Studies.

NOTE

1. One major company solves this problem by granting those on the top rung of their technical ladder the title of vice president.

REFERENCES

Allen, T.J. (1977). *Managing the flow of technology.* Cambridge, MA: MIT Press.

Allen, T.J., & Cohen, S. (1969). Information flow in R&D labs. *Administrative Science Quarterly, 14,* 12-19.

Allen, T.J., & Katz, R. (1986). The dual ladder: Motivational solution or managerial delusion? *R&D Management, 16,* 185-197.

Allen, T.J., & Katz, R. (1989). Managing engineers and scientists: Some new perspectives. In P. Evans et al. (Eds.), *Human Resource Management in International Firms* (pp. 191-199). London: Macmillan.

Allen, T.J., Katz, R., Grady, J.J., & Slavin, N. (1988). Project team aging and performance: The roles of project and functional managers. *R&D Management, 18,* 295-307.

Allen, T.J., Tushman, M.L., & Lee, D.M. (1979). Modes of technology transfer as a function of position in the R&D spectrum. *Academy of Management Journal, 22,* 694-708.

Badawy, M. (1988). Why managers fail? In R. Katz (Ed.), *Managing professionals in innovative organizations: A collection of readings* (pp. 162-169). Cambridge, MA: Ballinger.

Bailyn, L. (1980). *Living with technology: Issues in mid-career.* Cambridge, MA: MIT Press.

Dalton, G.W., Thompson, P.H., & Price, R.L. (1982). The four stages of professional careers: A new look at performance by professionals. In R. Katz (Ed.), *Career issues in human resource management* (pp. 129-153). Englewood Cliffs, NJ: Prentice-Hall.

Emmons, W.D. (1977). The dual ladder: The pioneering research approach. *Research Management,* 27-30.

Goldner, F.H., & Ritti, R.R. (1967). Professionalization as career immobility. *The American Journal of Sociology, 73,* 489-502.

Gouldner, A.W. (1957). Cosmopolitans and locals: Towards an analysis of latent social roles. *Administrative Science Quarterly, 2,* 281-306.

Graen, G., & Ginsburgh, S. (1977). Job resignation as a function of role orientation and leader acceptance. *Organizational Behavior and Human Performance, 19,* 1-17.

Hall, R. (1968). Professionalization and bureaucratization. *American Sociological Review, 33,* 92-100.

Kanter, R.M. (1977). *Men and Women of the Corporation.* New York: Basic Books.

Katz, R. (1988). Organizational socialization. In R. Katz (Ed.), *Managing professionals in innovation organizations: A collection of readings* (pp. 355-369). Cambridge, MA: Ballinger.

Katz, R., & Tushman, M.L. (1979). Communication patterns, project performance, and task characteristics: An empirical evaluation and integration in an R&D setting. *Organizational Behavior and Human Performance, 23,* 139-162.

Katz, R., & Tushman, M.L. (1981). An investigation into the managerial roles and career paths of gatekeepers and project supervisors in a major R&D facility. *R&D Management, 11,* 103-110.

Katz, R., & Tushman, M.L. (1983). A longitudinal study of the effects of boundary spanning supervision on turnover and promotion in research and development. *Academy of Management Journal, 26,* 437-456.

Kaufman, H.G. (1975). *Career management: A guide to combatting obsolescence.* New York: IEEE Press.

Kaufman, H.G. (1982). *Professionals in search of work: Coping with the stress of job loss.* New York: Wiley.

Kerr, S., & Von Glinow, M.A. (1977). Issues in the study of ¢professionals' in organizations: The case of scientists and engineers. *Organizational Behavior and Human Performance, 18,* 329-345.

Kornhauser, W. (1962). *Scientists in industry: Conflict and accommodation.* Berkeley, CA: University of California Press.

Kram, K.E. (1985). *Mentoring at work: Developmental relationships in organizational life.* Glenview, IL: Scott Foresman.

Pelz, D., & Andrews, F.M. (1966). *Scientists in Organizations.* New York: Wiley.

Raelin, J. (1985). *Clash of cultures.* Boston, MA: Harvard Business School Press.

Ritti, R.R. (1971). *The engineer in the industrial corporation.* New York: Columbia University Press.

Roth, L.M. (1988). A critical examination of the dual ladder approach to career advancement. In M.L. Tushman & W.L. Moore (Eds.), *Readings in the management of innovation* (pp. 275-292). Cambridge, MA: Ballinger.

Salancik, G.R., & Pfeffer, J. (1977). A social information processing approach to job attitudes and task design. *Administrative Science Quarterly, 22,* 427-456.

Schein, E.H. (1978). *Career Dynamics.* Reading, MA: Addison-Wesley.

Schein, E.H. (1988). How 'Career Anchors' hold executives to their career paths. In R. Katz (Ed.), *Managing professionals in innovative organization: A collection of readings* (pp. 487-497). Cambridge, MA: Ballinger.

Shepard, H.A. (1958, August). The dual hierarchy in research. *Research Management,* 177-187.

Steiner, G. (1988). The creative organization. In R. Katz (Ed.), *Managing professionals in innovative organizations: A collection of readings* (pp. 201-214). Cambridge, MA: Ballinger Press.

Tushman, M. (1979). Work characteristics and subunit communication structure: A contingency analysis. *Administrative Science Quarterly, 24,* 82-97.

Tushman, M.L., & Katz, R. (1980). External communication and project performance: An investigation into the role of gatekeepers. *Management Science, 28,* 1071-1085.

Von Glinow, M.A. (1988). *The new professionals: Managing today's high-tech employees.* Cambridge, MA: Ballinger.

Webber, R. (1976). Career problems of young managers. *California Management Review, 18,* 19-53.

Weick, K.E. (1969). *The social psychology of organizing.* Reading, MA: Addison-Wesley.

Whitley, R., & Frost, P. (1973). Task type and information transfer in a government research lab. *Human Relations, 25,* 537-550.

TRADITIONAL AND ENTREPRENEURIAL CAREER PATHS:

VARIATIONS AND COMMONALTIES

Robert A. Page, Jr., Gregory K. Stephens, and Angela Tripoli

Katz (1982) and Dalton and Thompson (1986) have conducted research on the career paths of engineering and technical personnel, typically in the context of large, formalized firms. They describe what we term a *traditional* career, where engineers can expect to begin with dull tasks and detail work, earn more challenging technical assignments as they prove themselves, and eventually head important projects and move into management responsibilities. The authors use different, but comparable career stage models to describe this process. Many engineers are comfortable with the constraints and long time horizons presented by this type of career within such a corporate setting.

This paper, however, focuses on a different breed of technical specialist—the entrepreneur—and the more risky and "discontinuous" career paths they prefer, whether in the context of established organizations or new ventures.

Human Resource Strategy in High Technology.
Advances in Global High-Technology Management, Volume 1, pages 151-175.

ISBN: 1-55938-489-1

For our purposes, the concept of "entrepreneur" will include anyone who defines their career primarily in terms of pursuing opportunities for creativity and innovation. These entrepreneurs do not value the rewards offered by more conventional career paths, and are willing to assume significant risks and sacrifices in order to pursue opportunities which demand more innovation and skill. The attitudes of these entrepreneurs, and the type of psychological contracts that govern the organizational relationships they develop with their employers, distinguish the entrepreneurial career path from the more traditional career path of the typical professional engineer. This paper explores the entrepreneurial career paths of technical specialists, paying special attention to the shifts in the psychological contracts between the engineers and their organizations at different stages of the entrepreneurial career. This analysis is based on data from two in-depth case studies, and also draws upon findings from the careers and high technology/innovation literature.

RESEARCH METHOD

The primary case study used to develop the analysis presented in this paper explored the career history and expectations of employees at a computer graphics firm, which, for the purposes of confidentiality, will be called Multi-Graphics, Inc. (MGI). MGI is a medium size high technology firm with annual sales of $100 million, and features product lines in computer aided design and computer aided training systems. The systems in both product lines are competitively differentiated primarily by their innovative technologies and superior performance capabilities.

The study lasted for more than a year, and involved in-depth interviews with all the members of the executive team, the senior engineering staff, and over 75% of the junior engineering staff. The interviews were relatively unstructured, adapting an "ethnographic" interview approach outlined by Spradley (1979) and Lofland (1976). Each interview began with a request for a description of the individual's career history. During this description interviewers probed for personal motivations and expectations. Once this initial stage was completed, the remainder of the interview focused on product histories and the process of technical innovation.

All interviewers were trained in qualitative research methods and had participated in at least one ethnographic study of an organization prior to this project. To minimize interviewer bias, two or more members of the research team conducted each interview. After the interview, notes were compared and interpretations of interviewee's statements were cross-checked. Each interview was transcribed within 48 hours to guard against memory loss and to avoid mental confusion with other interviews. The interviews ranged from one hour to three hours in length, with the typical interview lasting 90 minutes. In approximately one half of the cases, follow up interviews were scheduled for issue clarification or to solicit additional information. Most of the executive and senior engineering staff were interviewed several times. Additionally, three engineers on the junior level and one on the senior level were selected as key informants, and consulted on a regular basis during the interview and initial theory building stages of the project.

During initial theory building, whenever patterns of responses were identified, interviewees were asked for counter-examples, to avoid "exampling"—selectively developing a body of interview data which supports a given hypothesis (Glaser & Strauss, 1967). The selection procedures of MGI provided an excellent opportunity for collecting data and generating theory on entrepreneurial career paths, because MGI was staffed almost exclusively with entrepreneurial engineers. This company pursued a strategy of being technologic first movers in their industry, and actively recruited highly creative and innovative engineers from the best academic institutions across the country in pursuit of that goal. Most engineers at MGI reported that they were interested in pursuing entrepreneurial career paths, and joined the company primarily for that purpose. Consequently, the similarity among the patterns of response of these engineers regarding their career histories and expectations, coupled with the lack of significant contrasting examples, allowed a level of theoretical redundancy and diminished incremental learning from the interviews (Glaser & Strauss, 1967) to be attained relatively quickly concerning entrepreneurial career paths.

Kidder (1981) also describes the career paths, motivations, and expectations of highly creative and innovative engineers of the Eagle computer development group at Data General in narrative form in his book *The Soul of a New Machine.* His study was based on longitudinal observation and interviews of members of this development group, and his description is remarkably consistent with the case descriptions of the engineers at MGI. Consequently quotations from his book have been included occasionally to illustrate certain points in the theoretical analysis.

PSYCHOLOGICAL CONTRACTS

The psychological contract is defined here as an evolving set of mutual expectations, including, but not restricted to issues such as dependency and autonomy, human relationships, and individual and environmental change (Levinson, Price, Munden, Mandl, & Solley, 1963). While technical specialists do enter into a formal employment contract with their organizations, the majority of expectations regarding the rights and obligations of both parties remain informal, even unspoken (Gibson, 1966). These psychological contracts operate as powerful determinants of behavior in organizations (Schein, 1965). The organization and the technical specialist pursue an ongoing exchange relationship—or "reciprocation" process—wherein the parties carry out the psychological contract as each perceives it (Levinson, 1965). For both traditional and entrepreneurial specialists it is the psychological contract which binds them to their organizations. Individuals join, remain and make contributions to the organization because of their expectations concerning the probable rewards they will receive in exchange for their participation. These implicit contracts are dynamic, unfolding concomitantly with one's progression through different career stages. The traditional and the entrepreneurial technical specialists differ most clearly in terms of the content and the evolution of the psychological contracts which each maintains with their respective organizations.

THE CONVENTIONAL TECHNICAL CAREER PATH

The conventional career path of the technical specialist as described here involves three basic stages drawn from the research of Dalton and Thompson (1986) and Katz (1982): Apprenticeship, Independent Contributor, and Stabilization or Management. These approaches are compared in Table 1.

Table 1. Career Stage Models for Engineers

	Katz (1982)	*Dalton and Thompson (1986)*
Stage 1	**Socialization** Explore identity Establish roles Understand norms/expectations/ appropriate behaviors	**Apprenticeship** Directed by mentor Develop competence/responsibility/ reciprocity/communication networks
Stage 2	**Innovation** Focus on improving task performance Develop influence	**Independent Contributor** Independence Develop professional identity/expertise
Stage 3	**Stabilization** Overspecialization Career inertia/stagnation Decreased expectations Less performance focus	**Technical Expert** Narrow technical focus Either lacks managerial skill or rejects managerial career **or** **Mentor** Middle management responsibilities Leaves technical roots Mentoring/reciprocity
Stage 4		**Sponsor/Director** Top management responsibilities Provide direction Representation Selecting/sposoring/ developing key people

The first stage, referred to as apprenticeship (Dalton & Thompson, 1986) or socialization (Katz, 1982), deals with the training and orientation of the recently hired technical specialist. This stage is characterized by low responsibility and high dependence on others for guidance. The psychological contract between such new hires and their employers involves expectations such as eventual promotion and accrual of significant responsibilities in developing new technologies once the new hires prove themselves through consistent, quality work on repetitious, relatively dull tasks with low associated risk for the organization.

As technical specialists establish a reputation for dependability and competence, they become accepted as independent contributors (Dalton & Thompson, 1986) capable of quality innovation (Katz, 1982). In this stage, the specialist may become responsible for portions of important projects in the firm, and establish a reputation for technical expertise among peers and managers. They begin to desire greater independence, and may experience painful separation from Stage I managers and supervisors as they seek greater independence and autonomy. As independent contributors, the expectations of the psychological contract established during organizational entry can begin to be fulfilled.

Once the technical specialists have proven their worth as independent contributors, they face a major career choice—to remain focused on a technical area of expertise, which Katz (1982) refers to as the stage of stabilization, or to move away from that expertise into a management stage. The psychological contracts of those who move into management become fundamentally different from those who continue on as technical specialists. Individuals entering a management career path expect a greater leadership role in controlling the funding, staffing, and direction of technologic development. In turn, the organization expects they will assume an increasingly broad range of managerial and administrative responsibilities, even at the expense of remaining competent in their area of expertise, which Dalton and Thompson (1986) describe at length in the "mentoring" and "sponsorship" stages of their model.

These "traditional" career stage models have intuitive appeal because they seem representative of the majority of technical specialists, particularly those who progress into management. However, this general model does not delineate the distinctive career paths of individuals with differing career orientations. While the traditional career path may be representative of many technical specialists, it is proposed here that the traditional career path is not consistent with the attitudes, motivations, and career choices characteristic of entrepreneurs. In the following sections we compare and contrast the traditional career path with the career path of entrepreneurs whose desire for creativity and innovation takes precedence in managing their careers.

THE ENTREPRENEURIAL CAREER PATH

While the entrepreneurial career path features career stages similar to the traditional path (apprentice, independent contributor, and management or stabilization), the content of those stages, and the nature of the expectations involved in the psychological contracts typically negotiated between the technical specialists and their organizations, are quite different. Table 2 summarizes these differences.

Table 2. Entrepreneurial Career Stage Expectations

Entrepreneur	*Organization*
Stage 1: Apprenticeship/Socialization	
Challenging tasks	Top quality recruits
Innovative projects	Strategy of radical innovation
Superior leadership	Total commitment
Informal work environment	Low compensation
Stage 2: Independent Contributor/Innovator	
Gain respect of superiors and major customers	Self-selection Rites of passage
Develop innovations	Total commitment
Demonstrate autonomy/expertise	Entrepreneurship
Greater compensation/resources	Average salary
Stage 2: Redefinition of Management	
Rejoin traditional career path	Accept managerial responsibilities
Greater influence/resources	Develop managerial skills/abilities
Greater compensation	Loss of technical competence
Stage 2: Redefinition of Technical Specialist	
Lower expectations	Security
Limit entrepreneurial activities	Contractual relationship
Pursue outside interests	
Stage 2: Redefinition of Entrepreneur	
Continue attempts to innovate	Rapid promotion
Exit to start-up or competitor with/better compensation and resources	*or* Anticipate turnover/burnout
Stage 3: Stabilization/Technical Obsolescence	
Lower expectations	Project evaluation
Overspecialization	Mentoring
Rejoin traditional career	Management opportunities

Selection

In both cases, the profile of these entrepreneurial technical specialists begins in graduate school. While many future engineers are willing to accommodate the demands of a conventional corporate career, the entrepreneurial specialists (who generally classify themselves as "the best and the brightest") often have different aspirations. They recognize that most large business firms tend to follow the "conservative model of innovation," which Miller and Friesen (1982, p. 3) describe as follows:

> Basically, the model implies that innovation is not a natural state of affairs, that it must be encouraged by challenges and threats, and that it requires a particular type of structure and an effective information processing system to make conservative managers aware of the need for change.

From the start, these entrepreneurial types want more "action," and being bright, innovative, and self-assured, they expect to be able to find a job which promises to meet their expectations. They value the ability to pursue the leading edge of technology, and feel ready to assume responsibility for developing sophisticated innovations. They demand flexibility and informality from their organizations, and have little tolerance for any type of formalized practices or procedures—however important—which interfere with task performance. They are confident in their ability to generate significant radical innovation, and feel that expectations of founding their own firm, or developing and leading a major product line of their employer, are both realistic and realizable in the not too distant future.

These values of radical innovation are shared by technologically-driven entrepreneurial firms typically founded by other such engineers, or former academics. Often found in dynamic and hostile business environments, strategists describe these first movers as "entrepreneurial" and "prospector" firms (Ansoff & Stewart, 1967; Collins & Moore, 1970; Miles & Snow, 1978; Miller & Friesen, 1982). "Prospectors" define organizational competence and strategic advantage in terms of finding and exploiting new technologies, and the markets they create. While most organizations quickly "converge" on a dominant technologic design and move toward more incremental product innovation (Tushman & Romanelli, 1985), prospectors specifically avoid developing long-term commitments to a single type of technologic process (Miles & Snow, 1978). Instead, prospectors prefer a leadership position in introducing discontinuous changes, often in the form of radical product innovations, into the industry.

Perhaps what is most distinctive about these entrepreneurial firms is that their executives and senior engineers tend to view their companies in instrumental terms, often valuing the opportunity to explore innovative technologies more than maximizing financial returns (Miles & Snow, 1978; Perry & Page, 1986). They derive great personal and professional satisfaction as pioneers on the leading edge of technology, and are willing to sacrifice some profit to maintain high levels of research and development. Not incidental to the success of this mission is the ability of such firms to attract and retain technical specialists with similar entrepreneurial drive and commitment.

Stage 1: Apprenticeship

The entrepreneurial technical specialist enters apprenticeship with a personal set of expectations involving significant responsibility and technical opportunity, superior leadership, and informal work conditions. In return for meeting the expectations of the technical specialist, the organization demands total job commitment and expects the employee to be satisfied with relatively low financial rewards. This combined set of individual and organizational expectations comprises the initial psychological contract in the first stage.

Significant Responsibility/Technical Opportunity

Junior engineers at both MGI and Data General almost without exception claimed that the most important aspect of their job was the opportunity to work with "sexy" state of the art technologies, to begin their careers on projects that involved some substantive challenge and responsibility. For recently hired junior engineers in traditional career paths at larger, more conservative firms, these types of opportunities would be unprecedented. By promising this type of work, the best and the brightest graduates from the top schools are often attracted to smaller entrepreneurial, innovative firms. One new recruit summarized:

> I got an offer at IBM to work on a memory chip, to see what could be done about improving its performance. Here I got an offer to work on a major new machine, which was gonna be the backbone of company sales. I'd get to do computer design. It wasn't hard to make that choice. (Kidder, 1981, p. 62)

Engineers at both MGI and Data General also exhibited remarkable consensus with regard the intrinsic satisfaction of their work. They described a profound sense of "self-fulfillment," "satisfaction," and "accomplishment" from creating new technical products and seeing them function well. Note these comments of a Data General Engineer:

> Look, I don't have to get official recognition for anything I do. Ninety-eight percent of the thrill comes from knowing that the thing you designed works, and works almost the way you expected it would. If that happens, part of you is in that machine. (Kidder, 1981, p. 273)

However, the phenomenon described is more than satisfaction. Overcoming challenging technical problems with creativity and innovation seems to produce a lasting sense of euphoria, described by various Data General engineers as a "high," a "golden moment," "almost a chemical change," "the most incredible, soaring moment in my life" (Kidder, 1981).

Superior Leadership

Apprenticeship is tolerated by entrepreneurs only because of the respect they feel for their superiors. The opportunity to work with talented, creative, innovative scientists, often "gurus" in their area of technology, who share in their "technologic vision" and ultimately win their respect, is a powerful motivator for the apprentice entrepreneurial technical specialist. In fact, some junior engineers at MGI reported they chose that organization expressly for the opportunity to work with such senior people.

These senior engineers often seem driven by visions of an ultimate product capability far beyond the reach of current technologies. One such engineer, who founded his own company to pursue his vision, described his goals for Computer Aided Design (CAD) products:

> In the future, engineers will be able to sit down with one of our computers, and they will not only be able to design individual components on the screen, but measure and test them as well. Once they have made appropriate design changes in the components, they will be able to assemble them as a complete product, and test their performance as a system. In effect, they will be able to build a prototype and take it out for a spin without ever leaving the keyboard. Our system will revolutionize design.

This vision of the ultimate product—technologically infeasible for the short-term future—keeps technical designs "open" and continually pushes out the frontier of creative endeavor (Ansoff & Stewart, 1967; Perry & Sandholtz, 1988). New designs, quite predictably, never measure up to such stringent criteria. Engineers in the firm find this vision both intimidating and exhilarating. Kidder (1981) notes that part of the reason Data General engineers were willing to work so hard for so long, and under intense pressure to complete the project, was the vision of their manager, and the inspirational talks he gave periodically in which he shared his vision—the importance and "glorious" nature of their emerging technology—with his subordinates.

This sense of drive and exhilaration as talented people conquer worthy and important technological challenges is a powerful motivator. Speaking of the biotechnology industry, Kenneth Jarmolow, corporate director of research and development for Martin Marietta, comments:

> The better, smaller entrepreneurial companies have a quality about them that I've seen in no others, including our own—an extremely powerful entrepreneurship, more powerful for academic people than a sex attractant. (Johnson, 1983, p. 22)

Informal Environment

An emphasis on informality and flexibility includes such mundane aspects as work attire and relationships, but also extends to structural and procedural aspects of organizational life, such as flexible working hours, access to needed equipment and resources, and the minimization of red tape. Entrepreneurs highly value the qualities of informal and flexible working arrangements. One MGI engineer commented, "The pay is not the best, but the working environment is good—you can usually do your own thing at your own hours."

The emphasis on informality and flexibility extends to structural and procedural characteristics of the organization as well. Engineers at both Data General and MGI tended to become uncooperative, if not subversive, when they perceived that "red tape" was interfering with their ability to accomplish their tasks.

Commitment

Radical innovation is a demanding activity, requiring near total commitment on the part of the employee. Consequently long hours, seven day work weeks, and intense performance pressure are characteristic of this work environment. Competing loyalties to family, outside interests, friends, and so forth are expected to be sacrificed on the altar of achievement. Indeed, senior engineers at Data General used strong family ties or outside interests as a criteria for rejecting candidates when they were recruiting for their radically innovative development project (Kidder, 1981). Expectations such as these are often explicitly, and invariably implicitly communicated to potential new hires. Individuals unwilling to make the expected sacrifices typically move or are moved into more conventional career paths.

Compensation

Given such demanding work, financial compensation is remarkably conservative for newly employed technical specialists in the entrepreneurial career path. One MGI engineer notes:

> MGI bases its salary curve on the national one, a nice bell-shaped curve that is supposed to be objective. It is not fair when you realize that engineers of all abilities are included in the national statistics, while at MGI all are far above average. If there were a national salary curve of engineers at this skill level, MGI would be in a very uncomfortable position.

Data General offers similar compensation—conservative salaries with no overtime, regardless of the hours required. However, most of these entrepreneurial technical specialists openly admit that they "don't work for money" (Kidder, 1981). Instead, organizations such as these are able to attract highly qualified and talented individuals through their reputations as radically innovative firms offering unique opportunities for new recruits to handle significant responsibilities in leading edge technological development. For individuals entering the entrepreneurial career path, such a reputation for a firm is far more important than an above average salary.

Stage 2: Independent Innovators

The primary task of the entrepreneurial technical specialist in the second stage is to demonstrate competence in a chosen field. Once such technical specialists feel their competence has been sufficiently tried and tested, and that their ability to make independent contributions and to lead innovative development efforts has been established, their set of expectations undergoes a metamorphosis. Unlike their more conventional counterparts, however, the struggle for independence of the entrepreneurial technical specialist is not with their former supervisor or project head, but more often is with the firm itself, which continues to operate willingly or unwillingly under the initial psychological contract. For the individual, three categories of expectations are subject to significant change at this stage, including expectations with respect to status, organizational support, and rewards.

Status

Innovators expect that recognition from peers, senior employees in their company, and in sophisticated user firms will accompany their outstanding technical work and innovative accomplishments.

> For me, the compensation was the reaction of our scientific customers—they were really pleased—and just seeing that the thing worked. The company is a little clumsy in the way they handle recognition, though. [The president] has, to my knowledge, still never seen the thing. . . . In fact, for the grand opening I never received an invitation to go until I hinted pretty strongly that we might need some engineering support in case it broke down. . . . There was also an official write-up of the project, but it only mentioned two senior engineers who had minimal involvement. One was involved in the early days, getting the contract hammered out; the other helped trouble-shoot halfway through. I try not to let stuff like that bother me.

Likewise Kidder noted that a senior Data General engineer, after receiving a standing ovation for his work at a meeting of a society of computer professionals, described the acclaim as "the ultimate reward."

At both MGI and Data General, winning such recognition from senior managers was a sort of rite of passage, and not easily won. For example, the project review engineers at MGI were notorious for their ruthless dissection of proposed technologies, exposing weaknesses and probing for needed improvements. While such critiques became the basis for months of further research and design, it was a grueling process. In the words of one observer, "People were literally beaten up, if they didn't get their act together."

Organization Support

Recognition was typically reserved for the innovative and entrepreneurial junior engineers who emerged from MGI's "self-selection system" (Perry & Page, 1986). While junior engineers could work on leading edge technologies, and often were given large responsibilities at MGI, such tasks were still regarded as detail work—developing small pieces of someone else's project and technology. By design, the most outstanding engineers soon become weary of such assignments, and create their own niche developing something more interesting and worthwhile. The founder relates the following story to illustrate the value of such a self-selection system:

> There was a famous astronomer in England, whose work was so original and innovative that astronomers and professors from all over the world would come to work with him every summer. He would give each of them a relatively meaningless task, doing computations that would normally be assigned to some graduate student. Some professors were content to spend all summer doing such work. Others would quickly grow bored, and would come to him with a proposal of work that they felt would be more rewarding and mutually beneficial. His response was always, "Of course!"

If the engineer is any good, the founder maintains, he will soon return with his own proposal for a more challenging assignment. If the proposal makes sense the engineer is allowed to pursue it. If, on the other hand, the engineer continues to perform the mundane work, waiting for someone else to give him another assignment, then he is the "wrong type" for MGI. He will not be terminated; instead, he will continue to receive low-level assignments.

Perry and Page (1986) note that this job assignment strategy is a deliberate self-selection system, designed to ensure that only those engineers with the desired qualities of "academic " professionalism, initiative, and autonomy are retained. These engineers clearly enact their own place in the organization and in doing so develop their careers as they desire. The founder encourages and facilitates this process of enactment by rejecting the notion that organizations should carefully create niches and "fit" new engineers into the company, and insisting instead that those good enough to retain, carve their own niches. The following anecdote, related by a respected senior engineer, illustrates an instance of this strategy in action:

> A few years ago, [MGI's founder] came to me and asked if I needed a new recruit on any of my projects. I told him that we could maybe use someone to sweep the floors, but other than that, we had things covered. He said, "I've got a new engineer I want you to take on anyway. Find something for him to do." So I did. I put the kid to work testing circuit boards. Within a few weeks, he was coming to me with suggestions for re-designing the boards. That kid turned out to be one of the brightest, most creative engineers I've ever known. From then on, whenever [MGI's founder] brought me someone new, I didn't ask questions.

As these radically innovative types develop exciting new technical ideas in the process of creating their niche, they expect the organization to support their development efforts. Most conceptions of organizational strategic decision making are based on "induced" models where questions of resource allocation and new product development are decided at the top (Burgelman & Sayles, 1986). Engineers and other technical specialists working under the assumptions of the entrepreneurial career path, however, develop expectations around "autonomous" strategic models. Here the organization remains responsive to "autonomous strategic initiatives" of its lower-level engineers, and will even shift the organization's strategic domain to include a new technology that appears to be particularly promising (Burgelman & Sayles, 1986). Of course, these entrepreneurial career-oriented technical specialists have no doubt that their innovative ideas are particularly promising.

Thus the ultimate dream of many entrepreneurial engineers at MGI, and one embodied in Tom West of Data General, is to develop a project that shapes the organization. To be able to take a technologic vision, transform it into a product line, and have the organization shift and change to encompass it is the supreme accomplishment of an internal entrepreneur. Burgelman and Sayles (1986, p. 143) have noted this process in more conservative firms as well:

> More brilliant or perhaps more risk-prone executives seek to change the disposition of top management and get them to accept a new business field as legitimate for corporate development.

Commensurate Rewards

In this second career stage, independent contributors begin to expect rewards commensurate with the self-perceived magnitude of their contributions to the organization. These technical specialists anticipate rapid advancement, substantial financial rewards, and perhaps even equity in the organization. As one such engineer at MGI commented:

> When I left school, I made plans to be independently wealthy (regarded as a net worth of five hundred thousand to one million dollars) between the ages of 35 and 40. And if this company will not let me realize my plans, I will go elsewhere.

Most corporations, even of the "prospector" type, are unwilling, or unable to satisfy the latter two expectations of radically innovative technical specialists in the second stage of the entrepreneurial career. They tend to be unable or unwilling to provide such high levels of research and development support to such a broad number of engineers, and can not offer the equity positions and lucrative financial rewards their radically innovative employees expect. While the company may offer additional rewards in terms of status, recognition, promotion, and merit pay, they recognize that this may be too little, too late. Further, with promotions often come managerial responsibilities that these innovative types are not prepared or willing to handle that introduce even more frustration and incentive to exit the firm.

In short, the more competent, experienced, and innovative the entrepreneurial technical specialists perceive themselves as being, the greater the disparity between the expectations of the employee and the organization. As the second career stage progresses, the entrepreneurial technical specialists become progressively more unwilling to accept rewards, status, and opportunity that are not commensurate with their perceived competence and contributions to the firm. This tension is resolved through a redefinition of the psychological contract.

Redefining the Psychological Contract

The fact of scarce resources limits the organization's ability to fulfill the raised expectations of the technical specialist. As the organization fails to carry out the new terms of the entrepreneur's psychological contract, the entrepreneur commonly faces three options: (a) move into management; (b) remain with the firm, despite perceived inequities; or (c) move to another firm or found a start-up. The entry into one of these new roles marks an important transition in the entrepreneur's career. The transition involves a redefinition of the psychological contract and often considerable work role adjustment. Two theoretical conceptualizations from the literature on careers help to understand this process of redefining the psychological contract.

Nicholson (1984; Nicholson & West, 1988) posits four forms of work role adjustment that are important factors in career transitions. These modes of adjustment are characterized by the extent to which the individual's adjustment in the transition involves personal change or role change. The first, replication indicates a transition in which little or no adjustment occurs in either the individual or the role. Absorption implies significant change in the individual, but not in the work role. Adjustment to a career transition through determination indicates that significant change is effected primarily in the role. Finally, exploration as a mode of adjustment involves meaningful change in both the individual and the work role.

The modes of adjustment employed by the individual and encouraged or tolerated by the organization will clearly affect both the process and outcomes of the contract redefinition process. In less rigid organizations where role change may be an acceptable mode of adjustment, technical specialists with desires to move into a nontraditional entrepreneurial career path are likely to find the organization less resistant to redefinition of the psychological contract. In organizations with less accepting attitudes toward role change, "absorption" of organizational role definitions may be demanded with accompanying unwillingness to consider determination or exploration as modes of adjustment to the transition.

The choices which the individual entrepreneur will make at this stage and the nature of his or her adjustment may also reflect changing attitudes of the individual toward his or her career orientation. Individuals express varying orientations toward their careers. Furthermore, these orientations are likely to change and evolve over time, so that behaviors and attitudes exhibited in one career transition may or may not be congruent with those shown at another transition point. Several typologies of career orientations have emerged in the literature on careers (see, for example, Derr, 1986; Driver, 1979, 1980, 1982; Schein, 1978).

In an established and widely used conceptualization, Schein (1978) describes individual career orientations as "anchors" which fundamentally affect individual decision-making and behavior. This occupational self-concept, or career anchor has three components, including the (1) the individual's self-perceived talents and abilities, (2) self-perceived motives and needs, and (3) self-perceived attitudes and values. Further, he suggests that career anchors are not fundamental aspects of the person's personality, although over time they may become relatively stable. Rather, the individual's career anchor evolves through interaction between the person and the work environment, and is discovered through actual experience. Career anchors both determine and are determined by the experiences of the individual throughout the work life. The choices made and the contract redefinition experienced by the technical specialist after the second career stage will clearly be heavily influenced by the individual's career anchor. Table 3 summarizes these relationships.

Table 3. Redefining the Psychological Contract

Emergent Career Anchor	*Career Transition*	*Mode of Adjustment*
Technical/Functional Competence	Adjust Inducements/ Contributions Mix	Replication
Managerial	Move to Managerial Career Ladder	Absorption
Creativity	Pursue innovative niche within organization	Determination
	Exit/explore opportunities in "start up" or other organizations	Exploration

Three career anchors identified by Schein (1978) are especially relevant at this stage in the entrepreneurial career path. Individuals whose self-image, needs and desires are meshed with feelings of competence in a specific functional or technical discipline are associated with a "technical/functional" career anchor. People with a "managerial" career anchor may be technical specialists in the early stages of their careers, but they eventually come to regard specific technical or functional jobs as necessary interim activities leading to managerial responsibility. Finally, a "creativity" career anchor describes individuals with a dominant need to build or create something entirely their own, although they may also exhibit talent and ability in overlapping anchors such as managerial and technical/functional competence.

The Managerial Competence Anchor

Regardless of where technical specialists are employed, often the only path to the kinds of rewards, recognition, support, and influence they desire is through a managerial career track. However, the managerial career path, which typically begins as a product manager and culminates in the corporate executive suite, involves responsibilities that technical specialists often find repugnant. Responsibilities frequently mentioned as distasteful fall into three broad categories, including intergroup boundary spanning activities, managing power relationships and organizational politics, and becoming involved in formalized procedures and paperwork.

The attitudes of entrepreneurial technical specialists toward these managerial activities are a result of the socialization they experience in developing their technical expertise. In his case on biotechnology, Kania notes these comments of one corporate analyst:

> Academia places a clear premium on irrationality and antisocial behavior. Many of our best scientists have spent little time developing their interpersonal skills and can be very difficult to control. (1982 p. 17)

Because the education of a technical specialist is typically more in depth and specific to a particular field, they often advance and defend their opinions with respect to their field with a variety of arguments and facts, while undermining, if not ridiculing alternative positions. Consequently, they have a tendency to come across as unreasonable, patronizing, egocentric, stubborn, and nonresponsive (Burgelman & Sayles, 1986; Kenney, 1986). Hence, the entrepreneurial technical specialist is generally not well prepared or willing to assume the responsibilities of a manager.

The responsibilities of a manager demand political skills and extensive interpersonal networks. Entrepreneurial technical specialists must go from product champions to organizational diplomats (Roberts & Fusfeld, 1981). They must find ways to communicate with groups they formerly ignored or held in contempt, build relationships of openness and trust with people from whom they formerly withheld information, and recognize and accommodate the concerns and interests of groups about whom they care little and may have defied in the past (Page & Dyer, 1990).

Researchers consistently note that interpersonal networks—informal communication channels usually based on preexisting friendships with members of the other groups—prove critical in successful product development (Burgelman & Sayles, 1986; Delbecq & Weiss, 1988; Dyer & Page, 1988). Many MGI project managers admit that without interdepartmental friendships, many innovative projects, especially those involving radically new technologies, would never be brought to fruition. This creates dysfunctions for entrepreneurial engineers who tend to neglect such networks in their single-minded pursuit of technology.

Despite the reluctance of many technical specialists to become involved in the corollary functions of the managerial career path, some who move in this direction find their new activities consistent with their needs and desires. It is unlikely, however, to assume that many individuals with managerial competence as a career anchor will come from the ranks of entrepreneurial technical specialists. Movement into a managerial career path for the entrepreneurial technical specialist, is usually perceived by such engineers as an undesirable career direction and would be accompanied by an "absorption" mode of role adjustment. The entrepreneur adapts to the role demands of the organization by altering his or her own expectations.

The Technical/Functional Anchor

Ironically, technical specialists whose needs and desires revolve around their technical discipline may dislike the prospect of remaining involved in radically innovative projects and internal corporate ventures almost as much as moving into management. One venture manager commented:

> Be a "start-up" expert? I don't see that as a viable role for me personally. To be an entrepreneur means a lot of headaches; it means a lot of work. My average work day is 20 percent higher now than that of the average guy here. . . . I do not intend to do that for this company again . . . . it's also a total investment of the ego. Not everybody can do that. You give up your family. It's risky; you need tremendous confidence in yourself (Burgelman & Sayles, 1986, pp. 103-104).

A default option for many technical specialists is to remain with their firm, or move to a different firm, and adjust their expectations downward with respect to compensation, status, and technical opportunity. As one MGI project engineer said:

> The sad thing is, I'm tired of fighting ridiculous battles. Development is too hard to use so much vital time and energy fighting unnecessary battles. Quite frankly, I draw the same paycheck whether I bust my butt trying to do something challenging, innovative and risky, or sit around shuffling papers. Jack, who pioneered the display system technology here, left as a result of these pressures. He had other innovations he wanted to try, and couldn't see going through that kind of process again. So he left. There are a lot of battle scars, and too much pain. I never want to do anything like [the innovative project just completed] again. It's not worth it.

Many technical specialists choose to accept a secure paycheck, more normal working hours, and the opportunity to pursue outside interests in return for working on less challenging forms of innovation and foregoing the status, compensation and technical opportunity associated with radical innovation and the entrepreneurial career path.

As Dalton and Thompson (1986) argue, these individuals remain valuable independent contributors to the firm, and may in fact, form a necessary critical mass of solid contributing organizational citizens. Many may never move out of this second career stage, but others may find that mentoring incoming young professionals offers satisfaction and a sense of fulfillment that the firm is unable to provide for their technical efforts. This type of adjustment is characterized as "replication." It often follows a period of unsuccessful attempts to adjust to the role through "determination," where attempted role manipulation results in little substantive change to the role parameters in the direction desired by the individual. These individuals assess their existing organizational roles and their personal characteristics and needs, and decide that the costs of remaining "entrepreneurial" are too high. By abandoning their entrepreneurial goals, they avoid the risks, uncertainties and workload inherent in entrepreneurial pursuits. Adoption of such a new set of role assumptions and expectations creates a psychological contract similar to those characteristic of the more traditional career path.

The Creativity Anchor

A subset of technical specialists are unwilling to compromise their hopes and expectations for status, rewards, and the opportunity to influence the direction of their companies through technological innovations. For these individuals, moving into a managerial track or acquiescing to partial satisfaction of their dreams and expectations are both unacceptable options. These individuals are not satisfied with anything less than continuation of the entrepreneurial career path. However, a continuation of the entrepreneurial career path may represent a discontinuation of the traditional career path from the perspective of the organization. Two possibilities are available for these individuals. They can attempt to renegotiate their psychological contract with their organization, changing the willingness of their employer to meet their expectations. However, as has already been pointed out, most firms are unwilling or unable to meet the demands of all technical specialists desiring to continue in the entrepreneurial career path. Scarce resources, particularly with respect to compensation, rewards, and technological opportunity are a fact of life, especially in established organizations. Moreover, the more established the organization, the harder it will be for a technical specialist to materially influence the strategic direction of the firm.

The entrepreneur who successfully redefines the psychological contract with the organization and continues the entrepreneurial career path, does so by changing traditional role demands. This choice to continue the pursuit of increasingly challenging innovative projects within the firm represents a "determination" mode of role adjustment, since the individual refuses to accept personal change as an option. This type of role adjustment allows both the individual and the organization to make mutually beneficial changes in the psychological contract, and to develop a customized entrepreneurial role for the technical specialist.

The remaining option for the individual is to leave the firm, either moving to another established firm, or starting a new venture. This course of action is characteristic of the "exploration" mode of adjustment, where the individual pursues role change, as well as personal change. Moving to an already established firm offers a short term solution, since compensation, organization support, and mutual expectations may be negotiated to some extent. However, over the long term the individual is likely to encounter the same difficulties as before. Attempting a new venture or start-up, however, offers the potential for satisfying all of the expectations and desires of the entrepreneurial technical specialist. The possibilities for financial reward and equity positions in the firm are great, the ability of the individual to influence the strategic direction of the firm is assured, and the potential to achieve substantially increased status in the professional community is apparent. As Burgelman and Sayles (1986, p. 120) note:

> You must compare what the outside entrepreneur can have in terms of incentives and freedoms with what a corporation like United can give. The outside entrepreneur can get equity participation, capital gains, and particular ways of financing. He has short lines of communication, little overhead, and can engage in courses of action that are "not done by United," and so forth. The only advantage for the inside entrepreneur is the stable paycheck, and one can wonder whether that is an incentive at all for the real entrepreneur.

Of course, achieving these goals is dependent on the viability and success of the organization, but the confidence and cockiness of these entrepreneurial technical specialists makes admission of the possibility of failure, an unlikely factor in the decision to begin a new start-up. Further, founding a firm is the only real guarantee that an individual has that working conditions and targeted technologies will remain in line with personal expectations.

Stage 3: Stabilization

Within 10 to 15 years after graduate school, technical specialists oriented to work in fields of high technology become technically obsolete—they lose their ability to understand and pursue radical innovations because their specialized skills and/or knowledge base is no longer state of the art (Burgelman & Sayles, 1986; Maidique, 1980; Katz, 1982). Robert Noyce, inventor of the integrated circuit and co-founder and chairman of Intel, notes:

> Maybe you can do good technical work for ten years, if you work hard at it, but after that the younger guys are better prepared. It's a question of technical obsolescence, if you will. (quoted in Bylinsky, 1976, p. 16)

At this point technical specialists are no longer able to continue the cutting edge technically innovative work that kept them on the entrepreneurial career path. They must either move into managerial positions—where their experience and intuition will be valued and will form the basis for earning the respect of the technical specialists they manage—or they must settle for routine engineering and technical work in familiar specialities. For those who have taken the route of the entrepreneurial technical specialist and have successfully formed a new venture, the move into management begins with the first day of the new start-up, and the culmination of the move into management is a matter of degree rather than an either/or choice. As the founders or partners of a technically-oriented firm, often with substantial equity in the firm, established reputations for technological excellence, and an ongoing influence on the strategic direction of the firm, the move into management is less traumatic than it would have been at an earlier stage in their career.

These successful radical innovators fill critical management roles. They become the respected sponsors that they themselves admired and wished to emulate at the beginning of their careers. They allocate critical resources and control project development (Pearce and Page, 1990; Quinn, 1985). As the founders and top executives of their firms, these technical specialists now in stage four of their careers are deeply committed to a shared sense of corporate mission and technological vision, arising out of past successes in the development of radical innovations.

SUMMARY AND IMPLICATIONS

In conclusion, there are groups of employees, such as the entrepreneurial technical specialists whose attitudes, preferences, and career paths are not adequately understood through current career stage theory. This paper provides a description of the career paths of such individuals, and provides a contrast between the psychological contracts characteristic of such career paths and their more traditional counterparts.

During the first stage of their careers, entrepreneurs and their organizations tend to hold mutually acceptable and complementary expectations. However, as these entrepreneurial engineers become capable of more autonomous and innovative contributions to their organization, they tend to perceive a disparity between their expectations of higher levels of organizational support, recognition and compensation, and organizational expectations, which are inevitably lower. This perceived disparity begins a critical period of role adjustment during which the technical specialists and the organization redefine the terms of their psychological contract.

The nature of this redefinition will stem, in part, from the individual's career anchor preferences and expectations, which have been evolving throughout the individual's career. Initially, most of the entrepreneurs in the case studies were selected because of a clear career anchor in creativity. However, over time some of these engineers discovered evolving predispositions towards different career anchors, with accompanying expectations and role preferences. Consequently, when these engineers encountered perceived disparities in their psychological contracts with their organizations, a wide variety of adjustment strategies was observed, reflecting the wider variety of career anchors and preferences now present in the engineers.

Consistent with the profile of a creativity career anchor, entrepreneurial engineers attempt to redefine their psychological contracts through "determination" and "exploration" modes of adjustment. These forms of adjustment involve changing the work role, in this case to secure the organizational support, recognition and compensation necessary to continue entrepreneurial, innovative engineering activities. When such organizational support is not forthcoming, dominant creativity anchored engineers will tend to exit. However, a surprising number choose to remain, manifesting emergent career anchor preferences towards management or technical/functional roles. The expectations of the organization remain somewhat congruent with their own evolving expectations and they follow traditional career paths into management or technical specialization, integrating their evolving personal and role expectations and demands through "replication" or "absorption" modes of adjustment.

The examination of the alternative career paths of entrepreneurs suggests several implications both for research on career pathing and for actual career management. First, it is important for both individuals and organizations to recognize that shifting expectations and desires on the part of one or both parties are inevitable features of psychological contracts. These shifts occur in part because of changes in perceptions of equity as individuals pass through different career stages, and in part because the individuals themselves change and evolve by discovering new career preferences, and by distancing themselves from past preferences that they come to perceive as undesirable or too costly. These shifts create the need for redefinition of the psychological contract previously in force. Consequently, incorporating an evolving concept of the career orientation concept into career stage models is important. The cases analyzed here provide evidence that career anchors can strongly influence career decision making, and that the evolving and changing nature of the individual's understanding of his or her career anchor will likely be a recurring facet of most careers. The ways in which the psychological contract is defined and redefined, and the modes of adjustment employed at each stage transition will, in part, be a function of the evolving career orientation of the technical specialist.

The cases examined here affirm that the transitions between career stages are critical periods of adjustment for both individuals and the organizations that employ them. It is during these transitions that psychological contracts are tested and redefined. If expectations are not met, the terms of the relationship must be renegotiated. If the firm provides limited opportunities for continuing creative endeavors at one career stage, the following stage will likely be characterized by unsuccessful attempts at work role redefinition, subsequent low job involvement, and eventual exit of these creative and visionary people. Given the skills and creative abilities of entrepreneurial technical specialists, they tend to be highly mobile, and a large number will likely opt to leave the organization for more promising creative opportunities. It is clear that a continuing drain of such highly creative and motivated people would have severe consequences for organizations in rapidly changing and innovative fields. The decisions of entrepreneurial technical specialists to remain with their firms and to make significant contributions depend largely upon the extent to which organizations are willing to meet their expectations at each career stage. Systematic empirical investigation contributing to our understanding of the dynamics of psychological contracts and career pathing, the nature of career transitions, and the factors involved in adjustment may facilitate the organizational retention of key personnel.

Individuals and organizations can do much to ease transition passages and facilitate negotiation of new psychological contracts. Based on the discussion in this paper, several recommendations may be offered. First, and most important, is the necessity for an assessment of both the process and the content of the transition and contract redefinition. It is critical that individuals involved in career transitions become conscious of changes or previously unknown components of their own expectations, needs, career orientations, skills and desires. Such self-appraisal might be made more feasible through professional career counseling provided by the firm, and facilitation of this process would clearly have benefits for both the individual and the organization. In the absence of such counseling, individuals might well look to their peers, mentors, or personal relationships to help expand awareness of their evolving expectations, needs, and career orientations. Kram and Isabella (1985) have indicated that, while individuals' relationships with their mentors provide important feedback opportunities and insights into organizational influences, peers in the organization can be a significant source of information, especially with regard to attitudes of the sort shown to be important in this paper. Other important sources include subordinates, extraorganizational contacts and professional peers, and family.

While awareness of the evolving career orientation is a critical aspect of the transition and contract redefinition process, understanding appropriate and acceptable modes of adjustment is no less important. From the individual's perspective, carefully examining past experiences of individual's in similar situations, ascertaining the extent to which role definition versus personal change is considered acceptable, and seeking feedback after incremental efforts to institute role change or personal change are viable methods for easing this process. Further, should the need for a transfer to another organization become apparent, it is important that similar information be sought with respect to other organizations under consideration.

Relatively little research has explored the psychological contract, making an examination of changing expectations and needs with respect to the career orientations of individuals a fruitful place to begin to understand the dynamics of the evolving psychological contract. The comparison in this chapter of traditional and entrepreneurial technical specialists illustrates the viability and even the necessity of alternative career paths for some organizational employees. Both organizational career researchers and managers need to explore the nature of alternative career progressions, as well as the benefits and challenges posed by such alternative career paths.

REFERENCES

Ansoff, H.I., & Stewart, J.M. (1967). Strategies for a technology-based business. *Harvard Business Review, 45,* 71-83.

Burgelman, R.A., & Sayles, L.R. (1986). *Inside corporate innovation.* New York: Free Press.

Bylinsky, G. (1976). *The innovative millionaires.* New York: Charles Scribner's Sons.

Collins, O., & Moore, D.G. (1970). *The organization makers.* New York: Appleton, Century, and Crofts.

Dalton, G.W., & Thompson, P.H. (1986). *Novations: Strategies for career management.* Glenview, IL: Scott, Foresman.

Delbecq, A.L., & Weiss, J. (1988). The business culture of Silicon Valley: Is it a model for the future? In J. Hage (Ed.), *Futures of organizations* (pp. 123-141). Lexington, MA: Lexington Books.

Derr, C.B. (1986). *Managing the new careerists: The diverse career success orientations of today's workers.* San Francisco, CA: Jossey-Bass.

Driver, M.J. (1979). Career concepts and career management in organizations. In C.L. Cooper (Ed.), *Behavioral problems in organizations* (pp. 79-139). Englewood Cliffs, NJ: Prentice-Hall.

Driver, M.J. (1980). Career concepts and organizational change. In C.B. Derr (Ed.), *Work, family and the career: New frontiers in theory and research* (pp. 5-17). New York: Praeger.

Driver, M.J. (1982). Career concepts: A new approach to career research. In R. Katz (Ed.), *Career issues in human resource management* (pp. 23-32). Englewood Cliffs, NJ: Prentice-Hall.

Dyer, W.G., & Page, R.A., Jr. (1988). The politics of innovation. *Knowledge in Society, 1*(3), 24-44.

Gibson, R.O. (1966). Toward a conceptualization of absence behavior of personnel in organizations. *Administrative Science Quarterly, 11,* 107-133.

Glaser, B.G., & Strauss, A.L. (1967). *The discovery of grounded theory.* Chicago: Aldine.

Johnson, R.S. (1983). Martin Marietta draws three biotech firms to "consortium." *Genetice Engineering News, 3,* 22.

Kania, E.M. Jr. (1982). *Case study of Advanced Genetics Sciences, Inc.* Cambridge, MA: Harvard Business School Press.

Katz, R. (1982). Managing careers: The influence of job and group longevities. In R. Katz (Ed.), *Career issues in human resource management* (pp. 154-181). Englewood Cliffs, NJ: Prentice-Hall.

Kenney, M. (1986). *Biotechnology: The University Industrial Complex.* New Haven, CT: Yale University Press.

Kidder, T. (1981). *The soul of a new machine.* New York: Avon Books.

Kram, K.E., & Isabella, L.A. (1985). Mentoring alternatives: The role of peer relationships in career development. *Academy of Management Journal, 28,* 110-132.

Levinson, H. (1965). Reciprocation: The relationship between man and organization. *Administrative Science Quarterly, 9,* 370-390.

Levinson, H., Price, C.R., Munden, K.J., Mandl, H.J., & Solley, C.M. (1963). *Men, management, and mental health.* Cambridge, MA: Harvard University Press.

Lofland, J. (1976). *Doing social life.* New York: Wiley.

Maidique, M.A. (1980). Entrepreneurs, champions, and technological innovation. *Sloan Management Review, 21*(2), 59-76.

Miles, R.E., & Snow, C.C. (1978). *Organization strategy, structure and process.* New York: McGraw-Hill.

Miller, D., & Friesen, P.H. (1982). Innovation in conservative and entrepreneurial firms: Two models of strategic momentum. *Strategic Management Journal, 3,* 1-25.

Nicholson, N. (1984). A theory of work role transitions. *Administrative Science Quarterly, 29,* 172-191.

Nicholson, N., & West, M.A. (1988). *Managerial job change: Men and women in transition.* New York: Cambridge University Press.

Page, R.A., Jr., & Dyer, W.G. (1990). The politics of radical technical innovation. In L. Gomez-Mejia & M.W. Lawless (Eds.), *Organizational issues in high technology* (pp. 95-110). Greenwich, CT: JAI Press.

Pearce, J.L., & Page, R.A. Jr. (1990). Palace politics and radical innovation. *Journal of High Technology Management, 1*(2), 193-205.

Perry, L.T., & Page, R.A. Jr. (1986). *Why radical product innovation does not occur in large organizations: The intolerance factor.* Paper presented at the annual meeting of the Academy of Management, Chicago, IL.

Perry, L.T., & Sandholtz, K. (1988). A liberating form for radical product innovation. In U.E. Gattiker & L. Larwood (Eds.), *Managing technological development: Strategic and human resource issues* (pp. 9-31). New York: Walter de Gruyter.

Quinn, J.B. (1985). Managing innovation: Controlled chaos. *Harvard Business Review, 63*(3), 73-84.

Roberts, E.B., & Fusfeld, A.R. (1981). Staffing the innovative technology-based organization. *Sloan Management Review, 22*(3), 63-75.

Schein, E.H. (1965). *Organizational psychology.* Englewood Cliffs, NJ: Prentice-Hall.

Schein, E.H. (1978). *Career dynamics: Matching individual and organizational needs.* Reading, MA: Addison-Wesley.

Spradley, J.P. (1979). *The ethnographic interview.* New York: Holt, Rinehart & Winston.

Tushman, M.L., & Romanelli, E. (1985). Organizational evolution: A metamorphosis model of convergence and reorientation. *Research in Organizational Behavior, 7,* 171-222.

PART III

TRAINING AND DEVELOPMENT

FACING A FUTURE OF 'PERMANENT WHITE WATER': THE CHALLENGE OF TRAINING AND DEVELOPMENT IN HIGH-TECHNOLOGY ORGANIZATIONS

Joseph C. Bentley

"Training is everything," wrote Mark Twain, "The peach was once a bitter almond; the cauliflower is nothing but a cabbage with a college education" (1961, p. 23). In the chaotic world of high technology, training may in fact be everything. When the normal state of affairs is rapid, pervasive, continuing, relentless change, how can organizations hope to stay up with the ever changing demands of the environment—not to mention stay ahead—except by constant training and development?

Training and development not only represents one of the answers to the dilemmas of change in high technology, it also can be part of the problem. In order to offer training and development to managers and technical people

Human Resource Strategy in High Technology.
Advances in Global High-Technology Management, Volume 1, pages 179-201.

ISBN: 1-55938-489-1

in high technology organizations, someone must, in effect, "stop the world," and state, "We will design and implement our training and development program on the basis of these objectives and those outcomes." Unfortunately, this strategy, widely accepted as necessary in order to implement an effective training and development program, has at least two major flaws.

First, the goals and objectives of the training and development program may be connected to behaviors not valued by the "real" organization. As Brunsson observes, in his book, *The Organization of Hypocrisy* (Brunsson, 1989) all organizations operate at two different levels, usually referred to as the "formal" and the "informal." The formal organization is used for the purposes of demonstration or display to the outside world. Through the use of rituals, the formal organization presents itself as effective, ordered, and rational.

The informal organization, on the other hand, is the source of significant action in order to get important things done. It follows its own rules, policies and procedures. For example, in a recent interview, a senior personnel officer at Citicorp stated that "over the years we have hired a great many talented and aggressive people. We have in place, of course, a great many first-class management training courses that teach not only the technicalities of banking, but the importance of patience in changing organization and the virtues of effective communication and sensitivity" (Pascale & Athos, 1981, p. 170). The dilemma, however, is that the training programs were designed and developed in order to support the formal system, the "accepted" way of doing things. The 'reality' was quite different, according to the Citicorp executive:

> But the informal system rewards aggressiveness and verbal skill. The term we use around here is 'tiger'—and it is not a negative image. The attributes sought are naked ambition, tremendous flexibility, willingness to change, and an assertiveness kind of 'meeting macho' where you verbally punch your points home and spear those who disagree. Problems arise, however, when an organization can't expand fast enough to absorb that kind of aggressive energy. Then the tigers turn into sharks . . . all swimming around in a confined space . . . waiting. The quicker students who attend these courses on the niceties of management are watching out of the corner of their eye to see who the managers are who are going to survive and thrive. 'Interpersonal virtue' is no match for the powerful force of role models who consistently exemplify aggressiveness and hustle. (Pascale & Athos, 1981, pp. 170-171).

According to this scenario, being a *tiger* in Citicorp is the way to move onto the fast track. No formal training and development programs, however, would—or even could—set out to identify or offer the appropriate skills: naked ambition, tremendous flexibility plus a willingness to change, a "meeting macho" style where one is willing to punch and spear those who disagree. Training programs connected to the formal organization—as all must be in order to be approved—will quickly be seen by the tigers as not only irrelevant but potentially harmful. As Brunsson (1989) observes, training programs advocating formal system values may make people "narrower or stupider than they were before" (p. 16).

A second serious flaw in the traditional design of training and development programs in high-technology organizations is that, since the "real world of action" is changing so rapidly, any set of management skills or behaviors runs the risk of becoming rapidly outdated and therefore irrelevant. Basic management assumptions regarding such issues as authority, span-of-control, communication, strategic planning, leadership, motivation, and even training and development, when placed within the chaotic universe of constant change, will not hold.

"Most education trains us for the past," observed novelist Stegner (1962), "as most preparation for war readies us for the war just over" (p. 111). No amount of effort or planning is going to compensate for the fact that all of our knowledge is about the past, and all of our important management decisions are in the future. And among the few things that we know is the fact that increasingly, the future is going to be more and more unlike the past.

In high-technology organizations, training may, in fact, be "everything." But not training done in the old ways. The definition of training, the design and the delivery of training, the value of training—all are under the same relentless pressure of change as other organizational elements. Defining and designing effective programs of training in high-tech organizations is a daunting challenge for the future.

THE SUPERMARKET SHELVES

"The supermarket shelves have been rearranged," writes Don DeLillio, at the end of his novel, *White Noise* (1985):

> It happened one day without warning. There is agitation and panic in the aisles, dismay in the faces of older shoppers. They walk in the aisles, trying to figure out the pattern, discern the underlying logic, trying to remember where they'd seen the Cream of Wheat. They see no reason for it, first no sense in it. The scouring pads are with the hand soap now, the condiments are scattered. Men in Sansabelt slacks and bright knit shirts. Women with a powdered and fussy look, a self-conscious air, prepared for some anxious event. They turn into the wrong aisle, peer along the shelves, sometimes stop abruptly, causing other carts to run into them . . . there is a sense of wandering now, an aimless and haunted mood, sweet-tempered people taken to the edge. They scrutinize the small print on packages, wary of a second level of betrayal. The men scan for stamped dates, the women for ingredients. Many have trouble making out the words. . . . In the altered shelves . . . they try to work their way through the confusion (pp. 325-326).

The novelist is describing his vision of the future, and our reaction to it: "There is agitation and panic . . . They walk in the aisles, trying to figure out the pattern, discern the underlying logic. . . . There is a sense of wandering . . . sweet-tempered people taken to the edge" (pp. 325-326). With this supermarket metaphor, the safe, orderly world that we have known and depended upon is no more, DeLillio tells us. In its place is a new arrangement:

disorderly, unreasonable, and confusing.

While poets, writers, and artists traditionally have seen into the future more clearly than the rest of us, these days we have all become visionaries in one sense in that we are experiencing the effects of pervasive, massive, continuous change. Consider the reaction of managers trying to figure out just how it is that the shelves have been rearranged and what they are supposed to do about it

> I think managers recognize the need to change for a very simple reason: Ten years ago, when they asked a subordinate to do something, it was done. Now, when they ask a subordinate to do the same thing, the subordinate stares at them just a little bit longer, and thinks about whether to do it.
>
> I used to manage hardware people. That was easy. I knew just what they did and how to motivate them. Now I manage software people. I don't know what they do. They don't know what they do. Need me?
>
> Hell no, they don't need me. They don't even remember to pick up their pay checks. They're impossible to manage.
>
> We used to be a small, local, paternalistic company. We took good care of our people, related well to the community, and made a lot of money. We're growing and modernizing. We want our people to grow and develop. We want them to think about their jobs differently. There are a lot of opportunities in this company right now. But all they want us to do is to continue to take good care of them. They're indignant that we want them to change. What's wrong with them?
>
> Opportunity? For what? We haven't had a promotion in this plant in 18 months. With our operational cutbacks, I'm not sure any of us will ever again move up in this organization. We've always loved the industry and living in this area is great. But I think we're all becoming stagnant. At least we sure rationalize well about why it isn't our fault that the status quo isn't working as well as it used to. (Cummings, Mohrman, Mohrman, & Ledford, 1989, p. 2)

Largely due to the relentless change in technology as well as the speed with which it occurs, no segment of business has struggled more with the consequences of change than high-technology organizations. It is a truism to state that the operating world of high-tech firms has become one of constant change. Quantum leaps in technologies themselves have generated changes in all other elements of organization life: production and manufacturing, marketing, distribution, recruitment, training, rewards and recognition, and on and on.

In periods of rapid change, one favorite solution is training. Providing workers, managers, leaders with training in new methods, more advanced technologies or new skills and attitudes is, without question, an *answer*. For example in the late 1970s and early 1980s General Electric announced the opening of a new multimillion dollar training center designed to train over 3,000 employees annually in new technology applications ("For Your Information," 1980).

As Latham (1990) observes, for every "problem" faced by business organizations—faulty communication, unmotivated workforce, restricted rules, inadequate skills, lack of participation—one common denominator emerges: training (p. 5).

COMMON MODELS FOR TRAINING

The most common model for training is the Instructional Technology model (Goldstein, 1986; Laird, 1978; Latham, 1988; Wexley & Latham, 1981). This model emphasizes needs assessment, precisely controlled learning experiences designed to achieve preestablished learning objectives, importance of performance criteria, and the inclusion of an evaluation process with which to provide feedback to individuals and to the system (Goldstein & Gilliam, 1990). Figure 1 presents the model in more detail.

According to Goldstein and Gilliam (1990) the important elements of this training model are these.

- *Organizational Analysis:* Within the needs assessment requirement exists an important need to establish the macrolevel role of training. Included in the process are such things as organizational goals, climate for training, as well as the internal and external constraints present in the organizational environment. Goldstein and Gilliam (1990) propose that it is at this stage that a decision is made as to whether training is the appropriate strategy for the organization.

  A second phase of the needs assessment model is the determination of which tasks are required on the job, and the identification of the knowledge, skills, and abilities (KSAs) that are critical in learning to perform those tasks.

  Finally, a third part of the needs assessment process is a *person analysis,* a determination of those KSAs already present in the target group as well as a definition of those KSAs that are lacking.

  According to the model, once the process of needs assessment is completed, then the designer is ready to move into a determination of the instructional objectives.

- *Designing the Training Environment:* Included in this part of the model are the selection and/or design of the instructional programs, including a consideration of learning principles, type of educational experience, instructor selection, and so forth.

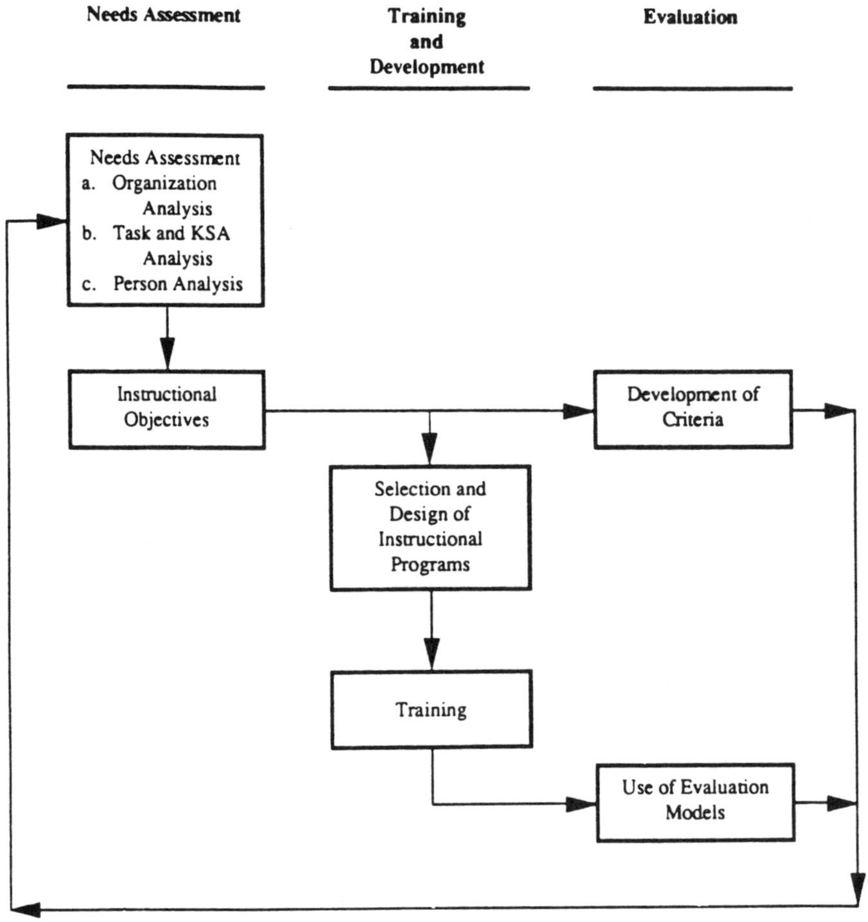

Source: Adapted from Goldstein (1986).

Figure 1. An Instructional Technology Model of Training

- *Evaluation:* The last set of elements in this instructional technology model is an evaluation of outcomes, including the development of success criteria (connected to the instructional objectives) and the application of evaluation methodology.

It seems clear that this Instructional Technology model contain several important, but unstated assumptions. Among these are:

1. A complex, semi-organized entity such as a large business organization can be reduced to sets of "facts": objective, widely-accepted conclusions such as a list of organizational goals, the determination of the climate for training, the identification of the internal and external constraints for training present in the organizational environment, a determination of those tasks required to be successful on the job, and a list of knowledge, skills and abilities (KSAs) critical in learning to perform those tasks (Goldstein, 1986).

2. Statements to the contrary notwithstanding, and in spite of the feedback loops present in the Instructional Technology model, this approach to training assumes a static or, at best, a slowly changing world in which skills learned in the present will continue to be appropriate for the foreseeable future. Since the Instructional Technology model begins with a diagnostic process consisting of identifying organizational goals, specifying important work tasks, selection of the relevant sets of knowledge, skills, and abilities, determination of the trainee's pretraining skills and aptitudes, and so forth, large amounts of time and resources must be invested in this process. Once completed, then, the preparation of instructional objectives, outcome criteria, selection and design of the instructional program can begin. The latter—the actual designing of the training itself as well as carrying it out—is dependent upon the previous diagnostic process. In the quasi-scientific spirit of this preassessment, months, even years could be invested. If such a diagnostic process were done well, resulting in valid information (Argyris, 1970) upon which to begin planning, it is not inconceivable that 12-18 months will have passed.

During these months, of course, the world has not slowed its frantic pace of change: turnover in the work force will have occurred; the markets will have made judgments about new products; strategic plans will have been tested; and so forth. Clearly, the Instructional Technology model of training does not fit into a world where the "supermarket shelves have been overturned." Since it is implicitly based upon assumptions that propose that "things will stay about the way that they are," it soon becomes outdated. As an example of the static world-view inherent in the Instructional Technology model, consider the training model found in Figure 2 (Laird, 1978).

3. The knowledge, skills and abilities (KSAs) necessary for accomplishing the important organizational tasks *can*, in fact, be taught and learned.

4. The KSAs to be taught as part of the training program will be relevant not only to the present task requirements, but can be expected to serve for as yet undiscovered or unknown tasks in the future.

Whether one can actually "know" what is happening in an organization, or whether the appropriate knowledge, skills or abilities can actually be taught are important issues; however, they fall outside of the purpose of this paper.

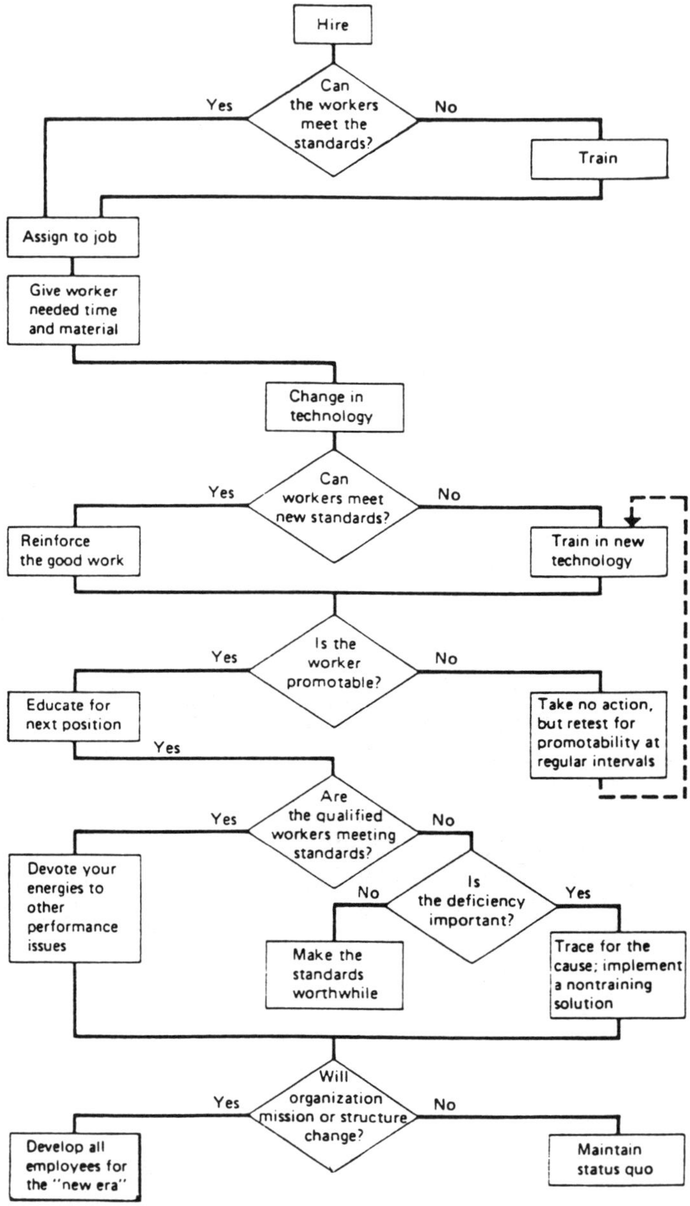

Source: Adapted from Laird (1978).

Figure 2. An Instructional Technology Model of Training with Built-In, Static World Assumption

Personally, I am not convinced that the "real organization" can ever be known, nor do I believe that "real learning" can occur in most if not all of the training programs designed and carried out within the Instructional Technology model. The second and fourth assumptions, however, are the most critical here: that under the Instructional Technology model the organization goals, mission, purposes, and so forth, can be expected to hold still long enough to develop training programs to teach the appropriate KSAs; and whether these KSAs can be made useful for the future. If, in fact, "the supermarket shelves have been rearranged," then a new and radically different approach to training in high-technology organizations is required.

The final question of Land's model (Figure 2)—Will organization mission or structure change?—reveals a bias toward the status quo and "business as usual." The alternative "maintain the status quo" as a response to this question is remarkable, given the upheaval of the past decade.

PERMANENT WHITE WATER

In 1989, a book by Peter Vaill appeared titled, *Managing as a Performing Art: New Ideas for a World of Chaotic Change* (Vaill, 1989). In Vaill's introduction, a powerful metaphor offers us a different perspective on the future. Vaill believes that his book "alerts managers to the context of chaotic change . . . upend[ing] much of conventional managerial thought." Managing a large corporation used to be like a pleasant boat trip down a calm, quiet river. The future, however, is going to be different. It will be full of rapids, whirlpools, eddies, and endless white water. The future as "permanent white water" will now be its most critical aspect.

Several years ago, as an example, Digital Equipment Corporation (DEC) greeted its thirtieth birthday in grand style. A 9-day celebration turned the majestic Queen Elizabeth II and the cruise ship Oceanic into floating hotels in Boston's harbor. Over 55,000 people celebrated with a party that cost $25 million.

Two years later DEC was nearly dead in the water. Unable to foresee the future slowdown in computer sales, unable to forecast the changing markets for traditional products, DEC embarked upon severe cost reduction, hiring freezes, and paycuts. Evidently, even the image of the Queen Elizabeth II was not enough to navigate the treacherous white waters of the late 1980s.

Another example of failing to negotiate the permanent white waters of the future is found in the book by Thomas J. Peters and Robert H. Waterman (1982), *In Search of Excellence,* a book that sold more than five million copies and appeared on the *New York Times* best seller list for over two years. Using a yardstick of six financial measures, Peters and Waterman identified 43 excellent companies. Five years later, two-thirds of these excellent companies had slipped off the pedestal. Some, such as Atari or National Semiconductor and six others, were in serious trouble. The others had either lost leadership or were in a seriously weakened position. Only 14 of the original 43 continued in the excellent category (Pascale, 1990).

In providing leadership and direction for large, complex organizations caught up in the agitation of permanent white water, what is it that leaders and managers are supposed to contribute? What is their "added value?" Among the many answers are these: They are supposed to find problems, "pathfind," solve problems, make decisions, and ensure that decisions are implemented (Argyris, 1970; Leavitt, 1986). While accomplishing these things, they will be (or should be) generating *valid information,* a rare commodity necessary in making *choices,* which in turn will generate *internal commitment* (Argyris, 1970). Defining leadership as "problem finding" and problem solving is a widely-accepted approach. But wait, argues Ackoff (1970). It makes no sense to talk about problems:

> Managers are not confronted with problems that are independent of each other but with dynamic situations that consist of complex systems of changing problems that interact with each other. I call such situations *messes.* Problems are abstractions extracted from messes by analysis . . . . Managers do not solve problems; they manage messes. (Ackoff, 1970, p. 95, emphasis in original)

If, as Ackoff proposes, the primary contribution of managers and leaders is to manage "messes," then *the* central question becomes: How do these managers and leaders learn how to manage messes? It seems abundantly clear that traditional models of training and development, relying as they do upon an Instructional Technology model, are not appropriate for preparing a manager or leader to be successful in a future full of "messes" and "permanent white water."

WHAT IS TO BE LEARNED

In considering the challenge of designing and implementing training and development programs for high technology companies facing a future of "permanent white water," another question emerges: What is to be learned? The answer, according to Campbell (1988) is a difficult and painstaking task, often avoided if at all possible: "It does not deal in high-tech bells and whistles and has little glamour associated with it." In the marketplace . . . it is a thankless task to argue that beating one's head against the question is far more important than using computerized instruction, having well appointed training facilities, or stacking a library with the latest videotapes" (p. 188). Indeed, for success in a future of permanent white water, the most important question may be what is to be learned? Here are several answers:

Becoming Intelligent

In his book, Gödel, Escher, Bach: An Eternal Golden Braid, Hofstadtler (1979) defines the intellectual abilities of an intelligent person. It seems important to begin with these:

> To respond to situations very flexibly;
> To take advantage of fortuitous circumstances;
> To make sense out of ambiguous or contradictory messages;
> To find similarities between situations despite differences which may separate them;
> To draw distinctions between situations despite similarities which may link them;
> To synthesize new concepts by taking old concepts and putting them together in new ways;
> To come up with ideas which are novel. (p. 21)

Taken seriously, these abilities have important implications for the design of training and development programs. If the ability to respond to situations in a flexible manner is important, if one should be able to make sense out of ambiguous or contradictory messages, then the training experience should provide situations in which such attitudes and skills can be developed.

Pathfinding

According to Leavitt (1986) there are three elements to the challenge of management/leadership: (1) pathfinding, (2) problem solving, and (3) implementation. Each of these parts asks for quite different sets of knowledge, skills, and abilities:

- *Pathfinding* is about getting the right questions rather than the right answers. It is about making problems rather than solving them. It is about mission, purpose, and vision.
- *Problem solving* is about analyzing, planning, and reasoning. It is an orderly, systematic approach to getting things done.
- *Implementing* is about doing, changing, and influencing. Here the key word is action.

Leavitt (1986) believes that the attitudes and abilities of Pathfinding are the most important in providing leadership for an organization in the midst of turbulent times. Yet, without doubt, they are the most neglected when training and management development programs are planned, designed, and carried out. "The pathfinding part of managing is the homeland of the dreamer, the innovation, the creator, the entrepreneur, and the charismatic leader" (pp. 10, 11). Imagine the consternation felt by a training director if they were asked to design a training program to develop such attitudes and abilities!

Vigilant Decision Making

There are four types of managers/leaders who will *not* be able to succeed in a future consisting of permanent white water, according to Janis (1989) in his recent book, *Crucial Decisions: Leadership in Policymaking and Crisis Management.* They are:

- Those who can't hack it, who overreact to the complexity of things;
- Those who can't say no, who cannot withstand the pleas of others;
- Those who are self-serving, whose personal needs for recognition or power overwhelm their judgment;
- Those who get discombobulated, those who become so anxious in the face of complexity and turbulence, that they cannot function. (pp. 216-222)

Those who can make it will be those who can develop and use their skills in *vigilant decision making.* Janis (1989) defines this set of abilities by proposing that the decisionmaker:

1. Surveys a wide range of objectives to be fulfilled, taking account of the multiplicity of values that are at stake;
2. Canvasses a wide range of alternative courses of action;
3. Intensively searches for new information relevant to evaluating the alternatives;
4. Correctly assimilates and takes account of new information of expert judgments to which he or she is exposed, even when the information or judgment does not support the course of action initially preferred;
5. Reconsiders the positive and negative consequences of alternatives originally regarded as unacceptable, before making a final choice;
6. Carefully examines the costs and risks of negative consequences, as well as positive consequences, that could flow from the alternative that is preferred;
7. Makes detailed provisions for implementing and monitoring the chosen course of action, with special attention to contingency plans that might be required if various known risks were to materialize (pp. 30-31).

Riding the Waves

Imagine the future as a powerful series of waves in which "managers and their organizations are confronting wave upon wave of change in the form of new technologies, markets . . . . competition, social relations, forms of organization and management, ideas and beliefs, and so on. *Wherever one looks, one sees a new wave coming"* (Morgan, 1988, p. xii, emphasis added). In using this powerful metaphor, Garth Morgan, in his book *Riding the Waves of Change* (1988), offers us a series of management challenges for the future. Following his metaphor, a successful leader for the future must be able to ride the waves of change. And what is it that one must know how to do in order to ride the waves successfully? Figure 3 presents Morgan's overview of the key managerial competencies.

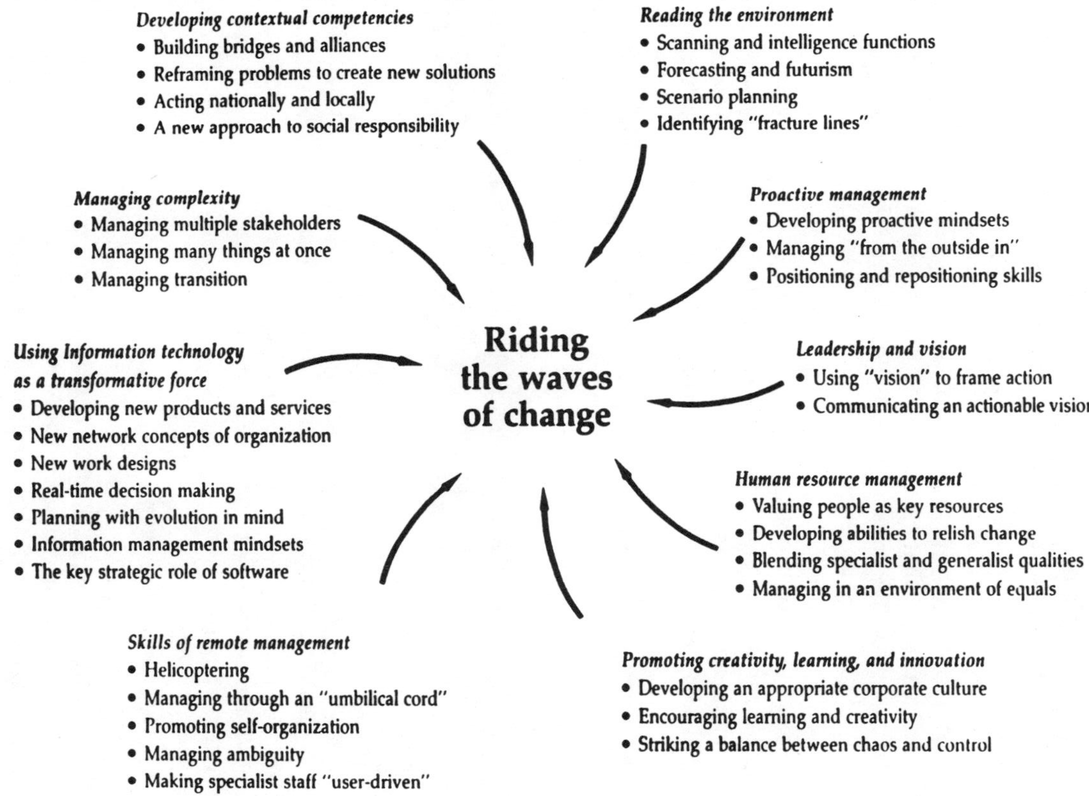

Source: Adapted from Morgan (1988).

Figure 3. An Overview of Some Emerging Managerial Competencies

Managing Paradox

Finally, in order to be successful while navigating through the white water of the future, one must be skillful in managing paradox, according to Robert Quinn and Kim S. Cameron in their book, *Paradox and Transformation* (1988). And just what, for managers and leaders, are paradoxes that need to be managed?

One answer is proposed by Cameron (1986), suggesting that organizational effectiveness is inherently paradoxical; that is, organizations must possess elements and attributes that are simultaneously contradictory or, under some circumstances, mutually exclusive. Sutton (1987) offers examples of paradoxes in his study of dying organizations:

1. keeping members from leaving so the work of disbanding could get accomplished, but at the same time moving members out of the system;
2. maintaining client and customer relationships in order to last out the disbanding process, while at the same time severing relationships with clients and customers;
3. keeping objects such as office equipment and inventories in order to maintain orderliness and minimal services, while at the same time getting rid of such objects through going-out-of-business sales;
4. taking away security of employment in order to help employees begin to search for alternative jobs, while at the same time maintaining security of employment so that employees would stay to help with the disbanding process and so that the best employees might even be reassigned within the broader organization.

Among the many suggestions offered in order to manage paradox successfully are these: tolerating ambiguity and uncertainty; balancing a rational either/or approach with one that accepts contradiction; an *and/and* approach (as in we need control *and* flexibility); reframing; and creating new and better organizational models.

What is to be Learned Revisited

When facing a future of permanent change, the central question returns: when preparing to design training and development programs for organizations eager to be successful in this turbulent future, What is to be learned? If the answers to the question include such knowledge and abilities as managing paradox, managing complexity, developing contextual competencies, learning pathfinding skills, and mastering vigilant decision making, then the model for developing programs of training and development must reflect these sets of skills and attitudes.

A PWW APPROACH TO TRAINING AND DEVELOPMENT

At the beginning of the 1980s, Peter Drucker (1980) issued a warning—perhaps a prophecy—that has since proved to be accurate: "The one certainty about the times ahead, the times in which managers will have to work and perform, is that they will be turbulent times" (p. 1).He continued:

> And in turbulent times, the first task of management is to make sure of an institution's capacity for survival, to make sure of its structural strengths and soundness, of its capacity to survive a blow, to adapt to sudden change, and to avail itself of new opportunities. (p. 1)

In this passage Drucker emphasizes the responsibility of management—senior managers and leaders—to ensure the survival of the organization. Yet there is abundant evidence that survival is not easily accomplished. Consider for a moment the interesting news that fully one third of all the companies occupying positions in the Fortune 500 in 1970 had, by 1983, disappeared (De Geus, 1988). Consider also the number of U.S. companies surviving until their fifth birthday (38%), their tenth (21%) and their twenty-fifth (7%) (Nystrom & Starbuck, 1984). Clearly survival, even in the "good old days" was problematical. While training and development can never serve as a panacea for all the problems faced by high technology companies in the future, it is critical; its contribution will be severely limited unless a more appropriate approach is developed and used. What follows are principles that could be used as a means of constructing a new and more relevant approach to training, training that will offer managers and leaders in the future better opportunities to navigate "permanent white water."

Unlearning Must Precede Learning

Rarely do trainers or educators in high-technology organizations begin at the beginning: acceptance of the fact that the cognitive maps that leaders carry around in their heads are inadequate and inappropriate. Human beings spend many years seeking answers to complicated and elusive problems. Once 'discovered,' those answers take on a quality of permanence. Gradually, in spite of evidence to the contrary, those answers become "the" answers. Since we become competent in any endeavor by making decisions about alternatives habitual to our unconscious, we are reluctant to open up our personal standard operating procedures for analysis, scrutiny, and, most difficult of all, revision. It is easier to believe that "we've met that problem and we know how to solve it. After all, we've faced that one dozens of times."

Unfortunately, the situations that gave rise to the tightly-held answers have changed so profoundly that the executive may possess answers for problems that no longer exist. And possessing what seem to be perfectly good answers often leaves no room for other possibilities.

Nystrom and Starbuck (1984) attribute the emergence of organizational crises to failure to unlearn the obsolete and irrelevant lessons from the past. Once organizations learn to address a certain type of problem "they encase their learning in programs and standard operating procedures" (p. 59) that then are transformed into routines. In some cases, this process contributes to an increasing level of efficiency, especially when the processes occur in a static, unchanging environment. When in the middle of "white water," however, clinging to routine often leads to unmanageable crises. "[T]op managers bolstered by recollections of past success, . . . misperceive events and rationalize their organizations' failures" (Nystrom & Starbuck, 1984, p. 59).

Unlearning is among our most difficult learning experiences, resisted with great energy. Therefore, in training programs designed to take seriously the challenges of turbulence, unlearning must appear early as part of the basic agreement between the instructors, and the learners. What follows must be a relentless examination of individual and group assumptions that are so jealously guarded. Wheelis (1973) has proposed suffering as a first step in his model of attitude and behavior change. While suffering may not be called for, a pronounced level of discomfort will be required, or well-established, obsolete cognitive maps will escape untouched.

The Learning Process Must Contain Those Attitudes, Values, or Skills that are Identified as Desireable Outcomes

If, as Ackoff insists, managers must know how to "manage messes," then managing messes must be part of the learning experience. If the future is to be full of uncertainty, ambiguity and change, then the learning process must contain ambiguity, uncertainty, and change. If successful managers must be able to come up with ideas that are novel, synthesize new concepts by taking old concepts apart and putting them back together in new ways, take advantage of fortuitous circumstances and so on (Hofstadtler, 1979), then these activities must be part of the learning process. If leaders are expected to be pathfinders (Leavitt, 1986) then ways must be found for them to learn to find paths through difficult and complex problems.

The Learning Experience Will be Suffused with Relevant, Accurate, Helpful Feedback

Most learners lack helpful and timely feedback. The Permanent White Water approach to learning must correct this flaw. It is a truism that without feedback, learners cannot improve their performance or acquire new attitudes. Yet many training programs limit their feedback to the content contained within the training design, and ignore the more important need: How will the learner manage and lead in difficult, stressful situations in the future?

While feedback is often used in many different ways, for a PWW model of management training, feedback should be (Swann, 1983):

1. delivered by others seen to be competent, and in a position for making an appropriate judgment;
2. experienced by the receiver as discrepant, at odds with one's own perception of one's self, yet not so farfetched that it is dismissed as ridiculous or absurd;
3. relevant to an important aspect of behavior;
4. delivered by a large number of people rather than an individual or a group;
5. delivered in ways that "up-end" people's view of themselves, create dissonance and discomfort, yet not in ways to create resistance or defensiveness; and
6. delivered by others who clearly support one's well-being and success.

Not Only Trying Harder, But Trying Differently

Among the "extra baggage" many of us carry through our lives are myths that sound like answers: "If it's worth doing, it's worth doing right" is an example. A "trying harder" myth is widespread. It tells us that for those who keep trying, who don't give up or give in, success is inevitable. The "trying harder" belief is usually connected to toughness, thus making it doubly difficult to back down or to give up. "When the going gets tough, the tough get going" is undoubtedly written on almost every high school dressing room wall.

Another approach makes sense. As Wheelis (1973) observes, trying differently may be as important as trying harder. Argyris (1990) describes seven errors identified by top managers as crucial in diminishing personal and organizational effectiveness. Examples of these errors are: "Actions intended to increase understanding and trust often produce misunderstanding and mistrust" (p. 6); "Blaming others or the system for poor decisions" (p. 7); and "The management team is often a myth" (p. 9).

One of Argyris' persistent problems is named "organizational inertia:" "The tried and proven ways of doing things dominate organizational life" (1990, p. 7). In other words, when faced with difficulties, the dominant strategy is to "do the same things that we've always done, but do it harder, or faster, or longer." Hornstein (1986) has called this the "rule of repeated action." When in doubt, "do what you did yesterday. If it isn't working, do it twice as hard, twice as fast and twice as carefully" (p. 12).

High-technology organizations, facing a future of "permanent white water" cannot afford the rule of repeated action. What must be built into training programs is the alternative: try differently. Among the skills and attitudes necessary for "trying differently" are reframing, unfreezing, creativity, and innovation. Management training that does not include creative decision making and problem solving will not be appropriate for an uncertain and ambiguous future world.

The Training Experience Must be Powerful Enough to Ensure Cognitive Reorganization as well as Increased Motivation for Constant Re-learning

A common mistake that many training designers commit is to seek to establish a comfortable, relaxed setting within which to carry out the learning experience. If change—real change—in attitudes, behavior, and especially in values is the purpose of the training, then situations in which people are relaxed or comfortable may contribute too a lack of change and growth. What seems to be important is what Bloom (1964) called a "powerful environment." Learning environments characterized by intensity, pervasiveness, and consistency have greater prospects of producing change in attitudes and behavior. Bloom introduces an interesting concept as he describes the benefits of a powerful environment: "the individual is completely engulfed in a situation which presses him from every angle toward a particular type of development or outcome" (p. 212). Thus, learning environments that increase the possibility of real change in attitudes, values, knowledge or behavior, should "engulf" the participant. Training programs resulting in total immersion may be called for.

The Learning That Prepares Leaders for Permanent White Water Must Happen Outside the Classroom

One of R.D. Laing's "Knots" (1970, p. 55) tells us that:

If I don't know I don't know
I think I know
If I don't know I know
I think I don't know

He follows this conundrum with another (p. 56, emphasis in original), more insidious and destructive.

There is something I don't know
that I am supposed to know.
I don't know *what* it is I don't know,
And yet I am supposed to know,

And I feel I look stupid
if I seem both not to know it
and not know *what* it is I don't know.
Therefore I pretend I know it.
This is nerve-racking
since I don't know what I must pretend to know.
Therefore I pretend to know everything.

Most managers and leaders in most organizations understand Laing's dilemma: "I don't know what I'm supposed to know, nor do I know how to find out what I'm supposed to know, so I pretend I already know." *Learning what one is supposed to know, and then learning it cannot occur only in a classroom setting.* It must also occur in the midst of "real world" experience.

In medicine, law, and education, learners learn what they need to know by doing it, accompanied by an experienced and caring mentor. It seems strange that in management—an extremely complicated process of perception, decision making, problem solving, and so forth—we persist in sending out neophytes without consistent guidance. Mentoring relationships, helping the learner learn while struggling with real issues, must become part of the training program for managers in high-technology organizations.

Training for the Future Must Move from Personal Competence to Systems Understanding

In the past, training programs have emphasized personal competence. Teaching the individual new skills has been the primary objective. The individual learner was the focus; the operating assumption was if you learn to do things differently, then you will be able to make the situation different.

It is now clear that we had it all backward. In the future, in addition to helping individuals become competent, a primary focus must be upon changing the system. The new teaching/learning assumption must be this: by changing the system, then individuals will also change.

While many system elements are important to understanding, only two will be considered here: rules and reward patterns.

Changing the Rules

Every social environment develops rules by which action occurs. Those who know the rules and plan by them become competent and are often seen as effective and successful. Changing the rules is a powerful way to change behavior. As De Geus (1988) points out, "we looked for ways to accelerate institutional learning. Curiously enough, we learned . . . that changing the rules . . . could be a spur to learning" (p. 72).

It is often the unwritten rules, not the written ones or formal ones, that create serious problems. When contradictions occur—as they always will—between the written and the unwritten rules, then instead of enhanced learning, effectiveness and morale diminish. Laing (1970) puts it most clearly:

> They are playing a game. They are playing at not playing a game. If I show them I see they are, I shall break the rules and they will punish me. I must play their game of not seeing I see the game. (p. 1)

Argyris (1990) has organized these nonproductive, unwritten rules into what he terms defensive routines. The logic of these defensive routines is contained in four propositions:

1. Craft messages that contain inconsistencies.
2. Act as if the messages are not inconsistent.
3. Make ambiguity and inconsistency in the message undiscussable.
4. Make the undiscussability of the undiscussable also undiscussable. (p. 27)

Effective training for managers and executives in high-technology organizations must focus upon the nonproductive elements of social systems. If participants could identify the undiscussable rules, and then explore why they are undiscussable, changing unproductive, unwritten rules would be underway.

Rewarding What is Important

All managers are taught the principle of appropriate reinforcement: observe the "correct" behavior (or elicit it) and then *reward it:* Among the most widely accepted principles of the behavioral sciences is that behavior that is rewarded tends to be repeated.

Problems occur, however, because in many if not most organizations, "correct" behaviors are often not even recognized let alone rewarded, while incorrect behaviors are. The reward subsystem is—or can be—a powerful force for shaping and directing action toward appropriate goals. Kilmann (1984) puts it in the strongest of terms: "If I could investigate only one aspect of an organization in order to predict its long-term success, I would choose the reward system" (p. xi). The important elements of the subsystem are these: (1) members feel that the reward system is fair; (2) they are, and will be, rewarded for high performance; and (3) the appraisal system provides them with regular and accurate information in order that they can improve. One important element of system understanding then, is the importance of alignment between rewards and performance.

While Learning is Important, Even More Important is Learning How to Learn

There is no doubt whatsoever, that in the future *all* managers and *all* leaders must be learners. The future will certainly depend on it. As more information is generated and more knowledge produced, the situation can only get worse. In fact, even today, it is impossible to know all that should be known. Before long, most of the "facts" we know will become outdated and irrelevant.

More important by far than learning facts or principles is the ability to know how to learn! Learning how to learn is the master set of skills required for the future. Learning how to learn requires the learner not just to engage successfully in the process of learning, but to gain an understanding of the learning *process itself.* While it is difficult to engage in a process and at the same time learn how the process could work better, that is what is implied by learning how to learn. An example of this complexity comes from Keller's (1989) book *Rude Awakening.* As General Motors began to plan their Saturn project, it occurred to them that repeating the old lessons once again would not work. Neil DeKoker, the director of Business Subsystems for the Saturn project, put it clearly: "We looked at all the disciplines to find out what we needed to do differently to become competitive. And the bottom line was that it wasn't just design, engineering, and manufacturing that needed to change, but *the entire business of running the business"* (Keller, 1989, p. 94). Not just running the business, but examining the business of running the business is an organizational equivalent of learning how to learn. It is what Argyris (1974) has called double-loop learning.

A CLOSING WORD

The orderly, predictable model of training known as Instructional Technology, will not serve those who, in the future, are charged with leadership in organization faced with "permanent white water." Instructional Technology, as an approach to training, assumes that the future will be enough like the past and the present to specify goals and objectives and design learning experiences to attack problems that can be specified, usually by some process of diagnosis or needs assessment.

An alternative view is that not only is the future unknown, it is unknowable. The only certainty is, for high technology organizations, that it will be filled with unsettling and unpredictable change. The supermarket shelves have indeed been overturned, and there is every reason to believe that they will be overturned again and again. An orderly, predictable, secure future existed perhaps, only in our memories, as we compared the past with our troubling present. Among all the uncertainties, one certainty emerges: that things will be more and more uncertain. Years ago, in *Walden,* Thoreau wrote, "For a man needs only to be turned around once with his eyes shut in this world to be lost. . . . Not till we are lost . . . do we begin to find ourselves" (1991, p. 139). Being lost then, may not be such a bad experience if, once lost, we can learn how to find ourselves.

REFERENCES

Ackoff, R. (1970). The future of operations research in past. *Journal of Operations Research Society, 30,* 93-104.

Argyris, C. (1970). *Intervention theory and method.* Reading, MA: Addison-Wesley.

Argyris, C. (1974). *Organizational learning: A theory of action perspective.* Reading, MA: Addison-Wesley.

Argyris, C. (1990). *Overcoming organization defenses: Facilitating organizational learning.* Boston, MA: Allyn & Bacon.

Bloom, B. (1964). *Stability and change in human characteristics.* New York: Wiley.

Brunsson, N. (1989). *The organization of hypocrisy: Talk, decision and action in organization.* Chichester, England: Wiley.

Cameron, K. (1986). A study of organizational effectiveness and its predictors. *Management Science, 32,* 87-112.

Campbell, J.P. (1988). Training design for performance improvement. In J.P. Campbell (Ed.), *Productivity in Organizations: Frontiers of Industrial and Organizational Psychology* (pp. 177-216). San Francisco, CA: Jossey-Bass.

Cummings, T.G., Mohrman, S.A., Mohrman, A.M., & Ledford, G.E. (1989). *Doing research: The case of organization design.* Los Angeles: University of Southern California.

De Geus, A. (1988, March-April). Planning on learning. *Harvard Business Review, pp. 74-82.*

De Lillio, D. (1985). *White noise.* New York: Viking Press.

Drucker, P. (1980). *Managing in turbulent times.* New York: Harper & Row.

For your information. (1980). *Training and Development Journal, 34,* 9.

Goldstein, I.L. (1986). *Training in organizations: Needs assessment, development and evaluation.* Pacific Grove, CA: Brooks-Cole.

Goldstein, I.L., & Gilliam, P. (1990). Training system issues in the year 2000. *American Psychologist, 45*(2), 134-143.

Hofstadtler, D. (1979). *Gödell, Escher, Bach: An eternal golden braid.* New York: Basic Books.

Hornstein, H. (1986). *Managerial courage.* New York: Wiley.

Janis, I. (1989). *Crucial decisions: Leadership in policymaking and crisis management.* New York: The Free Press.

Keller, M. (1989). *Rude awakening.* New York: Morrow.

Kilmann, R. (1984). *Beyond the quick fix.* San Francisco, CA: Jossey-Bass.

Laing, R.D. (1970). *Knots.* New York: Vintage Books.

Laird, D. (1978). *Approaches to training and development.* Reading, MA: Addison-Wesley.

Latham, G.P. (1988). Human resource training and development. *Annual Review of Psychology, 39,* 545-582.

Latham, G.P. (1990). *Developing and training an organization's human resources: A union perspective* (Work Place Topics, AFL-CIO Department of Economic Research), *1*(4), 5-13.

Leavitt, H. (1986). *Corporate pathfinders: Building vision and values into organizations.* Homewood, IL: Dow Jones-Irwin.

Morgan, G. (1988). *Riding the waves of change: Developing managerial competencies for a turbulent world.* San Francisco, CA: Jossey-Bass.

Nystrom, P., & Starbuck, W. (1984). To avoid organization crises. *Organizational Dynamics, 12*(1), 53-65.

Pascale, R. (1990). *Managing on the edge.* New York: Simon & Schuster.

Pascale, R., & Athos, A. (1981). *The art of Japanese management: Application for american executives.* New York: Simon & Schuster.

Peters, T. & Waterman, R. (1982). *In search of excellence.* New York: Harper & Row Publishers.

Quinn, R., & Cameron, K. (1988). *Paradox and transformation: Toward a theory of change in organization and management.* Cambridge, MA: Ballinger.

Stegner, W. (1962). *Wolf willow.* New York: Viking.

Sutton, R. (1987). *The process of organizational death: Struggling, disbanding, and reconnecting.* Working Paper, Department of Industrial Engineering, Stanford University.

Swann, W. (1983). Self-verification: Bringing social reality into harmony with the self. In J. Suls & A. Greenwald (Eds.), *Psychological Perspective on The Self* (Vol. 2., pp. 33-66). Hillsdale, NJ: Erlban.

Thoreau, H. (1991). *Walden, or a life in the woods.* New York: Vantage.

Twain, M. (1961). *Pudd'nhead Wilson.* New York: New American Library.

Vaill, P. (1989). *Managing as a performing art: New ideas for a world of chaotic change.* San Francisco, CA: Jossey-Bass.

Wexley, K.N., & Latham, G.P. (1981). *Developing and training human resources in organizations.* Glenview: IL: Scott Foresman.

Wheelis, A. (1973). *How people change.* New York: Harper & Row.

IDENTIFYING THE TRAINING NEEDS OF MANAGERS IN HIGH-TECHNOLOGY FIRMS:

A CASE STUDY

Lawrence S. Kleiman and Robert H. Faley

The proper development of an organization's managerial workforce cannot be overemphasized. This is especially true in high-technology firms where managers often supervise employees in jobs where the consequence of error is high. It is not surprising, then, that high-technology firms would be concerned about ensuring that their managers are equipped with the requisite abilities necessary to perform their jobs with the utmost proficiency.

One result of this is that managers in high-technology firms (like managers in many other firms) are asked to participate in a plethora of training programs designed to increase their overall effectiveness. These programs, which may be offered "in-house" or externally, are often designed to change managerial behavior in such general areas as judgement, leadership, sensitivity, initiative, and creativity (Byham, 1977).

Human Resource Strategy in High Technology.
Advances in Global High-Technology Management, Volume 1, pages 203-216.

ISBN: 1-55938-489-1

Unfortunately, many of these training programs are more likely to fail than succeed; they rarely create a positive and lasting influence on managerial performance back on the job (Byham, 1977; Zenger, 1980). For example, it is often a mistake to think that exposure to a brief training program on leadership can meaningfully change a manager's fundamental ways of behaving. This is especially true when the program itself is not designed/chosen as a response to a specific training need the manager may have.

We argue here that the effectiveness of many managerial training programs would be substantially enhanced if their focus were shifted from the general to the specific. That is, from focusing on general managerial traits and behaviors to focusing on the *specific* duties and responsibilities that are important components of a manager's job (Byham, 1977). We call this approach "performance-based" managerial training.

This view is not ours alone. Mitchell (1984), for one, strongly suggests that the content of managerial training programs should be expressed in "how-to" terms which pinpoint training needs specific to individuals and their jobs. Moreover, he recommends that these particular training needs should be identified based on the results of a thorough needs analysis. In a similar vein, Simpson (1975) notes that management training programs should be designed to remedy identifiable, specific, and solvable performance deficiencies experienced by managers on their jobs and not the more ambiguous deficiencies that traditional managerial training programs seem to focus on.

While the need for performance-based managerial training has been strongly recommended by various writers, training programs of this nature have been infrequently reported in the literature. Apparently, too few organizations are opting for this approach. This may suggest that many organizations too often choose programs based on a "seat-of-the-pants" strategy which fails to consider specific, task-based, managerial performance deficiencies. For example, some may choose programs because their competitors use them and others because the program's sales representative was persuasive.

The failure to consider management training programs that are task-based may also be due to perceptions that this approach cannot be adapted well to managerial jobs. For example, McCall, Morrison, and Hannan (1978), as well as Stewart (1968), believe that the specific tasks performed by managers cannot be identified, that activities involved in managerial jobs are so brief, varied, and fragmented that they defy reliable examination. This view may be further complicated by the perception that many existing job analysis techniques are inadequate for identifying the major activities involved in most managerial jobs (Mitchell, 1984).

PROJECT OVERVIEW

This paper is a comprehensive review and analysis of two related studies (see Kleiman and Biderman, 1989; Kleiman & Faley, 1990). Each was part of a larger project whose overall aim was to determine the appropriate content of performance-based training programs for managers employed in the Nuclear Power Division of a large public utility. Thus, a major part of both studies was to identify managerial task-based performance deficiencies which were correctable through training.

Grouping of Positions

Prior to commencing the project a decision was made to group managerial jobs into more homogeneous categories. The major aim of the grouping process was to identify subgroups of managerial positions where all managers within a particular subgroup performed a similar set of tasks and thus would potentially have a relatively more common set of training needs.

A two-stage grouping process was employed. First, all positions were categorized by management level (i.e., lower versus upper). Lower-level management positions were operationally defined as those in which incumbents supervised nonmanagement personnel; upper-level positions were occupied by individuals with at least one level of management reporting to them. The second stage of the grouping process involved statistically analyzing the positions within each management-level category in order to identify subgroupings which were homogenous with respect to the nature and frequency of the job tasks performed.

Sequence of Steps

Because of the decision to categorize jobs based on their level, two separate statistical analyses were performed. However, the methodology employed was similar in both analyses and consisted of the following series of six steps:

1. Develop a task inventory questionnaire which contained a "master list" of all tasks collectively performed by the managers.
2. Administer the questionnaire to a representative sample of managers.
3. Group the management positions based upon the similarity of tasks performed.
4. Determine the importance of each task within each job group.
5. Determine the extent to which performance deficiencies on each task were correctable through training.
6. Determine the training-need priority associated with each task for each subgroup of positions.

Definition of a Training Need

For the purpose of this study, a "training need" was defined as a job task which met each of the following criteria:

1. *It should be important to the manager's job.* Obviously, a performance-based training program should not focus on tasks of minor importance.
2. *It should be one in which a sizable number of the managers are having difficulty performing well.* Managers do not need to be trained on tasks that are already being successfully performed. "If it ain't broke, don't fix it."
3. *It should be one in which the manager's job performance can be improved through training.* As noted by Michalak and Yager (1979), the failure to perform a task satisfactorily may be due to a number of causes, the lack of proper training being only one. For example, a manager may be fully capable of performing a task, yet fail to perform it satisfactorily because of various obstacles which are unrelated to training (e.g., lack of adequate financial, material, human, or other resources).

STUDY 1: TRAINING NEEDS FOR LOWER-LEVEL MANAGERS

Step 1: Development of Task Inventory Questionnaire

A task inventory approach, similar to that used by Tornow and Pinto (1976), was used for the purpose of identifying a master list of all tasks collectively performed by lower-level managers. Task statements were derived from two sources—existing managerial task inventories found in the literature and information obtained from interviews with job incumbents within the Nuclear Power Division of the sponsoring firm.

The primary sources for the literature review were Curnow (1967), Hemphill (1959), and Tornow and Pinto (1976). The list of task statements compiled from the literature review was supplemented by the information obtained from individual interviews with a stratified sample of 33 lower-level managers. The interviews were necessary because it was our belief that many of the tasks performed by the managers in the present study were unique to the Nuclear Power industry, and thus would not have been identified in the general management literature.

Task statements were written according to the following guidelines:

1. The task had to be both, a job activity represented by a discrete unit of work with a discernable beginning and ending, and be executed by one individual.

2. It had to start with an "action verb" which stated the specific function performed by the manager.
3. The action verb had to be followed by the object of the verb, which could be people, data, or things.
4. The task statement had to specify the purpose of the task if the purpose were not inherently obvious.

The initial list of task statements was grouped into dimensions based upon their content. The statements and dimensions were then reviewed by a panel of five subject matter experts (SMEs). The SMEs were asked to comment on the clarity, degree of overlap, and level of specificity of the task statements, as well as the appropriateness of the dimensions.

Upon revision, the final draft of the task inventory consisted of 104 task statements which were grouped into eight dimensions. These dimensions are: (1) review and evaluation of subordinate work performance; (2) assistance of subordinates in their work performance; (3) human resource management and planning; (4) material/equipment procurement and planning; (5) establishment of a productive working environment; (6) interpretation and implementation of rules and regulations; (7) contact with people within the organization; and (8) contact with people outside the organization. The task statements associated with Dimension 1 are shown in Table 1.

Table 1. Study 1: Dimension 1 Task Statements

Edits drafts of subordinates' reports
Observes subordinates to determine work progress/procedural and safety adherence
Reviews work of subordinates and completes service ratings
Evaluates proposals submitted by subordinates
Examines technical specifications as to their appropriateness/applicability
Compares proposed procedures against guidelines, specifications, drawings, and so on
Questions subordinates about work progress/procedural and safety adherence
Inspects final work product
Inspects daily status reports, journals, logs, and so on
Reviews service ratings completed by subordinates for their employees
Administers and scores oral and written examinations as required for job advancement

Step 2: Administration of Questionnaire

A task inventory questionnaire was developed which listed each of the 104 task statements in random order. The randomization was done to mask three repeat items which were included in order to estimate intra-rater reliability. The questionnaire was administered in group sessions under standardized conditions to 137 of the organization's 160 lower-level managers. Respondents were asked to estimate, on a 5-point scale, the amount of time they spent on each task relative to all other tasks.

Two indices were employed to estimate the reliability of the time-spent ratings. The first was the correlation within each pair of repeat items. These correlations were .66, .80, and .85. Additionally, coefficient alpha was computed separately for each job dimension. These coefficients ranged from .70 to .89. Both sets of reliability coefficients are at conventionally acceptable levels.

Step 3: Identification of Job Groupings

Positions were grouped using the cluster procedure described by Ward and Hook (1963), followed by the Schoenfeldt (1970) affirmation process. The units of analysis were the 137 respondents. The dependent variables were the time-spent ratings (transformed to Z-scores), computed separately for each job dimension. The job dimension scores were derived by calculating the mean of the time-spent ratings within each dimension. Results of the cluster analysis revealed that five relatively independent job groupings existed based on the way job incumbents spend their time. For instance, Group 3 managers work most closely with their subordinates, spending relatively more time than other managers on supervisory-type tasks (such as assisting and evaluating their subordinates' work and establishing a productive work environment). Group 2 managers, on the other hand, are more involved in the interpretation of regulations than they are in supervisory-type tasks.

Step 4: Determining Task Importance

Two scales were used to measure task importance in the present study—"consequence of error" and "importance to overall job performance." The objective of the first scale was to provide information about the consequences of unsatisfactory task performance. The latter scale was designed to provide information on the extent to which performance on a task distinguished between inferior and superior workers. Each required that incumbents assess each task using a 6-point scale, where a 0 indicated the task was not part of the incumbent's job and a 5 indicated that the task was extremely important (i.e., had a high consequence of error and clearly distinguished between inferior and superior workers).

A questionnaire containing a list of the 8 dimensions, the 104 task statements, and the 2 task importance rating scales (as well as an additional scale to be described later) was administered to 38 SME representing the five job groupings revealed by the cluster analysis. Each SME rated the importance of each task within the particular job group for which they were considered experts. The number of respondents selected to represent each job grouping (13, 3, 6, 10 and 6 respondents from Groups 1-5, respectively) was roughly proportional to the number of positions within each grouping. The questionnaire was administered by mail. All 38 forms were completed and returned.

An index of the overall importance of each task was derived by computing the mean of the scores from the two importance scales. Thus, a task receiving a "consequence of error" rating of 2 and an "importance to overall job performance" rating of 4 would receive an importance index of 3.

The reliability of the importance ratings was estimated using the intraclass correlation coefficient, formula 3K (Shrout & Fleiss, 1979). A high intraclass coefficient indicates interrater agreement on task importance within a given job grouping. The reliability coefficients were .92, .53, .88, .88 and .84 for job groupings 1 through 5, respectively. (The relatively low coefficient for group 2 may be due to the small number of respondents ($n = 3$) within that job grouping.)

Step 5: Identifying Correctable Performance Deficiencies

The questionnaire described in Step 4 contained an additional rating scale—the "contribution of training" scale. This 5-point scale measured the extent to which training could improve the job performance levels of managers whose performance on a specific task was deficient. An index reflecting the extent to which performance on a specific task could be improved by training was derived by averaging the ratings made on this scale across all respondents regardless of job grouping.

The index was based on the average *across* job groupings because the impact the training had on task performance was not influenced by job-group differences. That is, the extent to which a training program could improve a manager's performance on a relevant task did not vary according to an individual's job responsibilities, but rather was primarily a function of the individual's initial level of proficiency. For example, someone entering a training program designed to improve interviewing skills who already had good interviewing skills would obviously tend to benefit less from the training program than someone entering with weak interviewing skills.

The resulting index ranged from 1 ("would not increase my proficiency") to 5 ("would greatly improve my proficiency"). A high score on the index indicated that the respondents perceived that their performance needed improvement and that improvement could be brought about through training. A low score indicated the reverse.

Step 6: The Determination of Training Need Priorities

A "training need priority" index for each task was calculated separately within each job grouping. This index was derived by multiplying the "overall importance" index (see Step 4) by the "correctable through training" index (see Step 5). Priority indices could range from 1 to 25 (1×1 to 5×5). High scores indicated that a task was both important and performance deficiencies on it were correctable through training.

A score of 6 was chosen as the cutoff for inclusion of a task as part of a training program for all incumbents within the job grouping where the index applied. This value was chosen because index values equal or greater indicated that some performance improvement was both necessary and possible.

Of the 104 tasks, 17 received a training priority score of 6 or above averaged across the five job groupings that resulted from the clustering procedure. These are:

1. Counsels subordinates on disciplinary problems
2. Advises subordinates on interpretation of guidelines and procedures
3. Informs subordinates of management's philosophies and goals
4. Communicates priorities to subordinates
5. Communicates rules and personnel procedures to subordinates
6. Sets goals for future performance
7. Examines technical specifications as to their appropriateness/applicability
8. Initiates improvements in subordinate's work methods
9. Provides initial guidelines for work assignments
10. Advises subordinates on direction to take with work problems
11. Seeks advice concerning the interpretation of regulations
12. Questions subordinates about work progress/procedural and safety adherence
13. Reviews work of subordinates and completes service ratings
14. Forecasts volume of work to be done for upcoming period
15. Forecasts future manpower needs
16. Sets priorities for total work group
17. Identifies morale problems via discussions with members of total work group

The following tasks received an average training priority score of less than 6, but had an average of 9 or above in one or more of the five job groupings:

1. Counsels subordinates on their professional development
2. Inspects final work product
3. Observes subordinates to determine work progress/procedural and safety adherence
4. Determines sequence of daily work activities for total work group
5. Discusses/answers grievances and EEO complaints with appropriate officials

STUDY 2: TRAINING NEEDS FOR UPPER-LEVEL MANAGERS

Step 1: Development of Task Inventory Questionnaire

The method used to assemble a master list of tasks germane to upper-level managerial positions was identical to the method used in Study 1: information gained from a literature review was supplemented with information gained from interviews with job incumbents. Interviews were conducted with 22 upper-level management personnel.

This procedure resulted in the identification of 106 mutually exclusive, non-overlapping tasks. As occurred in Study 1, these tasks were then grouped into 10 relatively independent dimensions based upon their content. These dimensions are: (1) fiscal responsibility, (2) public relations, (3) dealing with outside agencies, (4) human resource forecasting, (5) policymaking and planning, (6) personnel decisions, (7) compliance-related activities, (8) basic supervision, (9) training and development, and (10) serving as spokesperson for work group. The task statements associated with Dimension 5 are shown in Table 2.

The task inventory was independently reviewed by five SMEs using a structured review format. These experts met as a group on two occasions to

Table 2. Study 2: Dimension 5 Task Statements

Formulates statements of goals and objectives for total work group
Determines organization design for total work group
Develops strategies for achieving total work group objectives
Sets personnel policies within the division (e.g., recruitment, selection, promotion, transfer, termination, and discipline)
Initiates programs with regard to the training and development of human resources

discuss their recommendations regarding the clarity, overlap, and level of specificity of task statements, as well as the appropriateness of the dimensions. A consensus was reached on each of these points and the list of task statements was modified to incorporate suggested changes into a final draft.

Step 2: Administration of Questionnaire

A stratified sample of 33 upper-level managers was selected for the purpose of rating the 106 tasks on the four scales employed in Study 1: (1) time spent scale (how much time they spent on a task relative to the other tasks in a typical year); (2) consequence of error scale (the consequences of not performing the task well from a safety and/or efficiency viewpoint); (3) importance to overall job performance scale (the extent to which successful performance of the task was important to overall job performance); and (4) contribution of training scale (the extent to which training on the task would improve the respondent's job performance). The questionnaire was administered to groups of respondents under standardized conditions.

Step 3: Identification of Job Groupings

The purpose of this step was to categorize upper-level managerial positions into more meaningful job groupings. In Study 1, groupings were formed statistically by comparing the time-spent task profiles of each manager. That is, if managers had similar time-spent profiles (i.e., spent the majority of their time on similar tasks), their jobs would be classified in the same grouping. However, because of technical reasons, an analysis identical to that performed in Study 1 could not be done—there were too many variables relative to the size of the sample in Study 2.

A two-step process was implemented to reduce the number of variables. First, the number of variables was reduced from 106 to 10 by calculating the time-spent ratings for each *dimension* instead of each task. This was accomplished by averaging the time-spent ratings given by respondents for tasks within each dimension. Thus, the resulting time-spent profiles for each managerial position consisted of 10 scores, one for each dimension.

In the second step of the grouping process, these time-spent profiles were factor analyzed (varimax rotation). This further reduced the number of variables from 10 to three. The three factors were: (1) supervision (consisting of the dimensions: personnel decisions, basic supervision, training and development, and spokesperson); (2) policy making and planning (consisting of the dimensions: fiscal management, human resource forecasting, and policymaking); and (3) external contacts (consisting of the dimensions: public relations, dealing with outside agencies, and compliance-related activities). The factor loadings are reported in Table 3.

Table 3. Study 2: Factor Loadings Associated with Time-Spent Dimension Scores

	Factor Loadings[a]		
Dimensions	*Factor 1*	*Factor 2*	*Factor 3*
Fiscal Management	35	47	29
Public Relations	24	17	72
Outside Agencies	17	37	67
HR Forecasting	13	97	14
Policy Making	21	60	26
Compliance Activities	48	19	70
Basic Supervision	94	19	17
Training/Development	63	42	43
Spokesperson for Groupo	60	06	38

Note: [a]Decimal points have been omitted.

Finally, more homogeneous job groupings were identified using the hierarchical clustering technique described in Study 1. Factor scores were the units of analysis.

The clustering technique combined the factor-score profile of each respondent until the within-group variance dramatically increased (scree test). A two-group solution was obtained: all upper-level managerial positions were classified as belonging to one of two groups. The two groups differed primarily with respect to the amount of time they spent on Factors 1 and 3. Compared to Group 1 managers, those in Group 2 spent less time on supervisory tasks and more time on external activities. The intraclass reliability coefficient of the time spent ratings was .86 for Group 1 and .94 for Group 2. Both coefficients are at conventionally acceptable levels.

Step 4: Determining Task Importance

As in Study 1, the overall importance of each task was computed based on the mean rating it received on the two importance scales. The computations were done separately for each job grouping.

Step 5: Determining Correctable Performance Deficiencies

An index reflecting the extent to which performance could be improved by training was derived by averaging the ratings made on the "contribution of training" scale. The procedure is the same as that used in Study 1.

Step 6: Determining Training Need Priorities

A training need priority index for each task that was identical to the index used in Study 1 was calculated separately for each job grouping. This was accomplished by multiplying the "overall importance" index by the "correctable through training" index. The same cutoff employed in Study 1 (6 or greater) was used here.

Of the 106 tasks included on the questionnaire, 24 received an average priority index of 6 or greater within one or the other job grouping; 9 of the 24 were priority training needs for both groupings. These nine are:

- Formulates statements of goals and objectives for total work group
- Develops strategies for achieving total work group objectives
- Forecasts quantity and type of work to be done within total work group
- Answers regulatory agency questions concerning alleged violations
- Recommends corrective action to be taken with regard to regulatory violations
- Communicates priorities to subordinates
- Sets goals for subordinates
- Identifies morale problems within total work group
- Discusses EEO complaints and union grievances with appropriate officials

Tasks for which only Group 1 managers needed training are:

- Devises means of reducing operating costs
- Resolves conflicts between own subordinates and those of other work groups
- Counsels subordinates on their professional development

Tasks for which only Group 2 managers needed training are:

- Assesses inventory of talent within organization
- Determines organizational design (relationships, responsibilities, structure, and so on) for the total work group
- Seeks advice concerning interpretation of compliance regulations
- Communicates work standards expected of subordinates
- Gives recognition for subordinates' good work
- Completes performance evaluations for subordinates
- Inspects final product or outcome of subordinates' work
- Discusses progress/problems of total work group with superiors
- Discusses total work group's status reports and needs at meetings (e.g., staff, coordinator, and so on)

- Advises subordinates on the interpretation of technical guidelines and procedures
- Advises subordinates on direction to take with work problems
- Counsels subordinates regarding disciplinary problem

IMPLEMENTATION

The sponsoring agency initially used the findings of the two studies to evaluate the content of its existing management training programs. By carefully analyzing these programs the agency was able to determine which important tasks (i.e., those with high training need priority indices) were not currently incorporated in their existing training program curriculum. Plans are now underway to develop performance-based training modules that focus on those specific tasks.

As a first step in this undertaking, the agency plans to group related tasks that would "fit" into the same training module. For example, several upper-level management tasks with high training need priorities dealt with planning and goal setting and could be combined into a single module. However, a more detailed analysis of each task must be conducted before a complete module can be developed.

This analysis should provide both a step-by-step breakdown of how each task should be performed as well as information on the knowledge, skills, and abilities (KSAs) related to the effective performance of each task. Where a task is complex, it may be best to use an algorithm task-analysis approach to "capture" the problem solving/decision making process used in performing the task (see Zemke & Kramlinger [1982] for an excellent discussion of various algorithmic procedures). Traditional job analysis approaches are available for isolating the requisite KSAs. This step has already been completed for the lower-level management positions (see Kleiman & Biderman, 1989).

There are a variety of acceptable training techniques that could be used for improving managerial performance on tasks where deficiencies exist and training provides a meaningful solution. While a decision has not yet been reached regarding the specific training technique that will be used, it is our recommendation that the agency use behavioral modeling (see Goldstein & Sorcher, 1974). We recommend behavioral modeling because this technique requires both that the correct way of performing each task be demonstrated to trainees and that they practice it themselves. Moreover, detailed feedback is provided throughout the practice sessions.

CONCLUSION

The purpose of this paper was to point out the feasibility and advantages of taking a performance-based approach to the development of managerial training programs. While the results reported are unique to the particular organization studied, the method could be adopted by many other interested organizations.

We believe this approach is an effective, straightforward one which gets right to the heart of the matter—finding out how to improve managerial job performance. There have been some, however, who view this approach as radical. For example, when the sponsoring agency was presented with the results and recommendations noted above, one administrator stated in bewilderment: "You mean that we should actually teach our upper-level managers how to do their jobs? No, I don't think that would work here." We think it would. In fact, we think it is the most effective way to significantly improve managerial job performance.

REFERENCES

Byham, W.C. (1977, April). Changing supervisory and management behavior: Part 1. Problems with conventional training. *Training and Development Journal,* pp. 3-6.

Curnow, G.R. (1967). *The dimensions of executive work in the U.S. Federal Career Service: A factor analytic study.* Ann Arbor, MI: University Microfilms, Inc.

Goldstein, A.P., & Sorcher, M. (1974). *Changing supervisor behavior.* New York: Pergamon Press.

Hemphill, J.K. (1959). Job descriptions for executives. *Harvard Business Review, 37*(5), 55-67.

Kleiman, L.S., & Biderman, M. (1989). Job analysis for managerial selection: A guidelines-based approach. *Journal of Business and Psychology, 3*(3), 353-360.

Kleiman, L.S., & Faley, R.H. (1990). Determining training needs for upper-level management positions: A performance-based approach. *Journal of Management Systems.*

McCall, M.W., Morrison, A.M., & Hannan, R.L. (1978). *Studies of managerial work: Results and methods* (Tech. Rep. No. 9). Greensboro, NC: Center for Creative Leadership.

Michalak, D.F., & Yager, E.G. (1979). *Making the training process work.* New York: Harper & Row.

Mitchell, W.S. (1984, October). Professional management training needs analysis. *Training and Development Journal,* pp. 68-70.

Schoenfeldt, L.F. (1970). *Affirmation program.* Unpublished computer program, University of Georgia.

Shrout, P.E., & Fleiss, J.L. (1979). Intraclass correlations: Uses in assessing rater reliability. *Psychological Bulletin, 86*(2), 420-428.

Simpson, K. (1975). Management development: Full-spectrum training. *Training and Development Journal, 3,* 3-7.

Stewart, R. (1968). Management education and our knowledge of managers' jobs. *International Social Science Journal, 20,* 77-89.

Tornow, W.W., & Pinto, P.R. (1976). The development of a managerial job taxonomy: A system for describing, classifying, and evaluating executive positions. *Journal of Applied Psychology, 61,* 410-418.

Ward, J.H., & Hook, M.E. (1963). Application of a hierarchial grouping procedure to a problem of grouping profiles. *Education and Psychological Measurement, 23,* 69-81.

Zemke, R., & Kramlinger, T. (1982). *Figuring things out: A trainer's guide to needs and task analysis.* Reading, MA: Addison-Wesley.

Zenger, J. (1980, December). The painful turnabout in training. *Training and Development Journal,* pp. 36-49.

TECHNOLOGICAL CHANGE: ITS EFFECTS ON THE TRAINING AND PERFORMANCE OF OLDER EMPLOYEES

Robert C. Liden and Susan M. Adams

Results of studies examining the association between employee age and performance ratings are certainly among the more inconsistent reported in the organizational literature with correlations ranging from strong and positive to significantly negative (McEvoy & Cascio, 1989; Rhodes, 1983; Waldman & Avolio, 1986). Several moderating variables, such as job type and type of rating, have explained a small portion of the variability in age-performance rating correlations across studies (e.g., McEvoy & Cascio, 1989), but most of the wide variation in correlations across studies remains unexplained. One ignored yet potentially powerful moderating variable is technological change introduced into the workplace. The purpose of the current paper is to present a conceptual model outlining the way in which technological change may moderate the association between employee age and both actual and rated performance. In addition, research on training designed to prepare older employees for technologically changed jobs is reviewed. This topic appears to

Human Resource Strategy in High Technology.
Advances in Global High-Technology Management, Volume 1, pages 217-234.

ISBN: 1-55938-489-1

be of great practical as well as theoretical significance given projections for an older labor force and greater incidence of technological changes in the workplace.

The average age of labor force participants is increasing yearly and at the same time more and more jobs are being affected by technological changes (Bracker & Pearson, 1986; Sparrow & Davies, 1988). Considering these two changes simultaneously suggests that the jobs of many older employees will be changed technologically, implying that the jobs will be redesigned. Job redesign often involves changes in skill and ability requirements necessary for acceptable job performance (Rumberger, 1981). This raises critical questions such as: What are older employees' attitudes concerning technological change? How well do older employees respond to job redesign in terms of job performance? When their skills become obsolete, how well do older employees respond to training? Will supervisors be willing to give older employees an opportunity to be retrained? How accurately do supervisors assess attitudes and job performance of older employees who are faced with changes in their jobs?

These and other questions are addressed with the aid of a theoretical model presented in Figure 1. The model is based on the assumption that technological change typically results in job redesign which in turn often alters the skill and ability requirements needed for acceptable job performance. Given the change in skill and ability requirements, the job incumbent's success in developing new abilities and learning the new requisite skills becomes critical if job performance standards are to be maintained. Acknowledging the possibility of age stereotyping and/or biased perceptions of older employees, rated performance is treated separately from actual performance.

Although employee turnover is beyond the scope of this paper, the model indicates that at any stage in the process, the job incumbent may voluntarily or involuntarily withdraw from the organization. Voluntary withdrawal of older employees often translates as a decision to opt for early retirement. Variables such as financial security, interest and importance of leisure activities, relative interest and importance of the job, a spouse's decision to retire, and an employer's provision of training and promotion opportunities have all been shown to influence retirement decisions (Beehr, 1986; Robinson, Coberly, & Paul, 1985). Involuntary turnover may occur when physical and/or mental decline is judged to be too severe for training or retraining to successfully return employees to acceptable performance levels. However, because research evidence indicates that for most jobs such decline is rare, involuntary turnover among older employees is most likely caused by unfair discrimination.

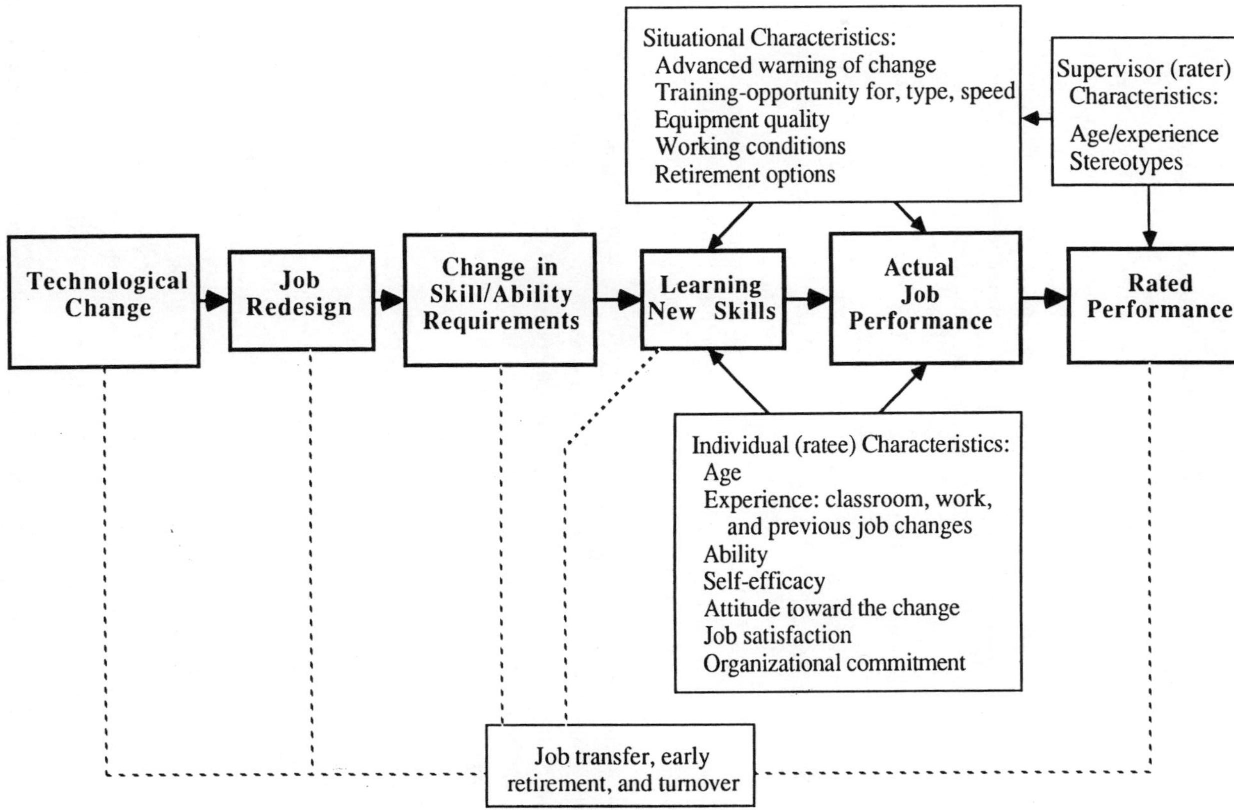

Figure 1. The Effects of Technological Change on Actual and Rated Performance

TECHNOLOGICAL CHANGE

Technological changes will be the rule rather than the exception in future decades, and in nearly all cases these changes will lead to the redesign of affected jobs. The degree of change can vary from slight, such as when an old process is simply speeded up by equipment improvements, to a complete change in the way jobs are executed. Research has shown type of technology to be related to employee perceptions of job characteristics (e.g., Hackman, Pearce, & Wolfe, 1978; Rousseau, 1977; Roznowski & Hulin, 1985). For example, Rousseau (1977) examined the association between technology, as defined by Thompson (1967) and employee perceptions of job characteristics. Her results showed that employees working on jobs utilizing long-linked (i.e., mass) technologies rated their jobs as being the least complex relative to the job characteristics perceptions of employees working on jobs categorized as using mediating or job order technologies. Similarly, Pierce (1984), found that "preprogrammed system-controlled technology" routinized jobs as reflected in employees' low ratings for job characteristics on the Job Diagnostic Survey (JDS) (Hackman & Oldham, 1976). Thus, an association between type of technology and job characteristics consistently appears in the literature.

Given that type of technology is directly related to job characteristics, it follows that technological change typically results in job characteristics changes. The first step in determining the effect that a given technology will have on job characteristics is to develop a taxonomy of technologies. Characteristics of different types of technologies (e.g., Collins, Hage, & Hull, 1988) can then be assessed in terms of job redesign that accompanies each type of technology.

JOB REDESIGN

In most cases, technological change will cause the characteristics of affected jobs to become either more or less complex. For example, Hackman et al. (1978) reported that the introduction of a computerized storage system for bank records resulted in some jobs becoming more complex than before the change, some less complex, and some essentially unchanged. In this and nearly all other job design studies (e.g., Griffin, 1981; Hackman & Oldham, 1976; cf. Loher, Noe, Moeller, & Fitzgerald, 1985) job characteristics were measured exclusively from the job incumbents' perspective using either the JDS (Hackman & Oldham, 1976) or the Job Characteristics Inventory (JCI) (Sims, Szilagyi, & Keller, 1976).

Recently, it has been stressed that these measures are overly narrow and ignore many other potentially critical aspects of jobs, such as physical demands (e.g., Campion & Thayer, 1985; Stone & Gueutal, 1985; Taber, Beehr, & Walsh,

1985; Zaccaro & Stone, 1988). It has also been argued that employee perceptions of job characteristics should be combined with assessment of job design from other sources such as supervisors (e.g., Campion, 1988, 1989; Campion & Thayer, 1985; Taber et al., 1985). A more comprehensive assessment of job characteristics is especially useful when attempting to identify the physical and mental skill and ability requirements for jobs (cf. Campion & Thayer, 1985).

Campion (1989) has suggested that job characteristics associated with job redesign should be assessed not only from the motivational perspective as in Hackman and Oldham's Job Diagnostic Survey, but also from mechanistic, biological, and perceptual/motor approaches. This is needed because the motivational approach captures mental aspects of jobs with its focus on job complexity, but ignores physical dimensions of jobs. Thus, a job that is redesigned considering only the motivational aspects of job characteristics may result in a job design that is inefficient (mechanistic approach), physically overdemanding (biological approach), or suffering from inadequate working conditions (perceptual/motor approach). Assessing both physical and mental characteristics of jobs is especially important when considering individual differences, such as employee age or sex, that may be related to specific mental and physical skills and abilities.

CHANGES IN SKILL AND ABILITY REQUIREMENTS

Considering physical as well as mental implications of job redesign enables the direct comparison of job demands with the mental and physical skills and abilities of people who are to fill the jobs in question. When current employees, who could handle all job demands prior to job redesign , no longer possess the skills and abilities necessary for completing tasks and responsibilities, obsolescence has occurred (Fossum, Arvey, Paradise, & Robbins, 1986). The size and nature of the gap between existing and needed skills and abilities determines whether new employees must be hired to fill the redesigned jobs, or if job incumbents can be successfully trained to handle the new job demands.

When deciding that current employees can be trained, careful assessment of the discrepancies between needed mental and physical skills and abilities and those actually possessed by employees may also be used in determining specific training needs (e.g., Goldstein, 1986). A thorough job analysis followed by training needs analysis has been recommended for making such assessments (Sterns, 1986; Sterns & Alexander, 1987). In addition, comparisons between skill and ability requirements for redesigned jobs and the skills and abilities that are most likely to show declines with age would indicate areas of special training needs and/or the need for further job design (Doering, Rhodes, & Schuster, 1983). The latter point is important to consider: Although not a focus of this paper, alterations in job design may reduce the amount of training necessary to return employees to acceptable performance levels.

When technological change results in jobs that are more complex, and further redesign alone is not adequate for assuring satisfactory task performance, training becomes critical. For example, a study involving 1,308 copy machine service engineers demonstrated that training mediated the relation between age and job performance (Sparrow & Davies, 1988). Although no job performance differences occurred in the 26-35 age group as a result of training, in the 36-45 and 46-55 age groups, those who had received training performed at significantly higher levels than those who had not been trained. In the following section, research from psychology, sociology, and gerontology that relates to the training of older employees is discussed.

LEARNING NEW SKILLS

Learning any new skill requires one to have the prerequisite skills and motivation coupled with the appropriate training for that individual (Goldstein, 1986). The acquisition of prerequisite skills, the motivation to learn, and the appropriateness of a training program may be affected by age-related issues. Age-related factors that are relevant to new skills acquisition may be categorized as being individually- or organizationally-based. An individual's existing knowledge, skills, and abilities (KSAs), self-efficacy, and attitudes may facilitate or hamper the learning of new skills. Additionally, the failure of an organization to provide adaptive training for older workers may decrease the success of retraining efforts that are focused on the acquisition of specific skills and abilities.

Individual Influences on Learning

The individual brings specific levels of knowledge, skills, and abilities (KSAs), self-efficacy, and attitudes to the learning experience. Older workers may be at a disadvantage entering the learning process and subject to learning problems if new job requirements or prerequisite skills for training include increased demands on abilities that tend to decline with age, such as reaction time or eyesight (e.g., Droege, Crambert, & Henkin, 1963; Ellis & Oscar-Berman, 1989).

Even though the decline in abilities is individual-specific, some generalities can be made. Physical characteristics such as strength and coordination of physical efforts are affected by the age-related decrement of bones, muscles, senses, neural impulses, and the cardiovascular system (e.g. Weg, 1975). Disease and disease-related complications increase. However, studies that do find gradual declines with age report only a 10 to 15% loss in neural capabilities

(Weg, 1975). Also, general cognitive abilities, as measured with standardized tests, decline somewhat with age (Avolio & Waldman, 1990; Salthouse, 1988), but these results appear to be due in part to generally lower educational levels and lower test-taking familiarity of older employees (Salthouse, 1988; cf. Kimmel, 1988). Older individuals may have been away from the learning environment longer than younger employees and hence, lack the necessary prerequisite learning skills necessary to succeed in the learning experience offered by the organization.

Research concerning declines in mental abilities has shown that semantic memory (e.g., immediate recall of letters) declines with age while episodic memory (e.g., recall of specifics embedded in events) may not (Smith & Fullerton, 1981). Similarly, fluid intelligence (i.e., one's capacity) decreases with age while crystallized intelligence (i.e., one's actual working aptitude) often increases with age (Botwinick, 1977; Schaie, 1990). Surprisingly, dramatic declines in abilities may not actually result in performance declines on well-maintained tasks. Even though some abilities may decline, adaptive behaviors can be adopted to compensate for those declines. Salthouse (1984) found that older typists adapted to the loss of tapping rates and reaction times by planning ahead and using larger eye-hand spans.

In assessing overall declines in skills and abilities with age, it is critical to point out that physical and mental abilities do not decline significantly with age prior to the time that most labor force participants retire (Brousseau, 1981; Coates & Kirby, 1982). In addition, in many gerontology studies, "older" has been defined as individuals 65 years old or older, a majority of whom have retired from employment. Given the small magnitude of decline in abilities that are directly attributable to age, other factors, some of which are correlated or interact with age, have been explored for their possible affect on learning abilities of older employees.

Work experience appears to be one of those factors. Although age and experience are often correlated (Avolio, Waldman, & McDaniel, 1990; Sterns & Alexander, 1987), previous experience with changes, in general (cf. Pinder & Schroeder, 1987), and with the types of new requisite skills may directly affect learning abilities. Changes involving new technologies will favor the worker who has had more similar educational or work experiences, whether he/she is young or old, by preparing the individual for prerequisite or requisite skills. Similarly, a person who is experienced in coping with changes, in general, may adapt to changes in job requirements more easily.

One factor often cited as a reason for poorer training performance and higher training turnover rates of older workers is a lack of recent classroom experiences (Smith, 1981). The lack of this prerequisite skill of classroom learning can hamper the learning process. Just as classroom experience may effect performance in a training program, work experience may affect the success that an older employee has in adjusting to job redesign: In both classroom and job settings, experience significantly contributes to ones self-efficacy, or confidence in mastering the new task (Bandura, 1977; Gist, 1987), making avoidance and flight behaviors less likely (Gist, 1987).

Self-efficacy has been defined as "one's belief in one's capability to perform a task" (Gist, 1987). The process of acquiring self-efficacy involves an individual assessment of capabilities from informational cues. These cues can come from four primary sources (Bandura, 1982). In order of influence, they are (1) enactive mastery; (2) vicarious experience; (3) verbal persuasion; and (4) emotional arousal. Enactive mastery and vicarious experience can be provided by organizations through training.

Enactive mastery comes from performance accomplishments. Training programs for older employees may require alterations to accommodate slower reflexes, reduced agility, dexterity, hearing, and vision (Doering et al., 1983). Research suggests that adjustments to training programs should also be made to accommodate the generally slower cognitive processing of older individuals to increase the likelihood of accomplishments in training (Gist, Rosen, & Schwoerer, 1988; Smith, 1981). Other individual differences, such as experience, ability, and self-esteem are also related to the enactive mastery component of self-efficacy.

The second most influential informational cue affecting self-efficacy is vicarious experience. Behavioral modeling as a training technique provides both enactive mastery and vicarious experience (Gist, 1987). Given projected trends in demography, many role models should be available as older individuals are expected to occupy a significant portion of the labor force (cf. Zenger & Lawrence, 1989).

Just as self-efficacy can motivate an older worker to learn new skills for a job redesign, the worker's perception of the new situation is a potentially important motivator. Older workers, as opposed to younger workers, appear less motivated to perform tasks that they perceive as being unmeaningful (Walsh, 1975). Similarly, Fossum et al. (1986) described two economically based circumstances under which older workers might be less motivated to want retraining. First, older workers may feel that newly developed KSAs will not be used long enough after training to be adequately rewarded. Second, sufficient increases in compensation may not be offered to older workers at the top of their salary ranges. There is some evidence to confirm the notion that older workers are less motivated to enter the training process. Studies by Blanchard-Fields and her colleagues (Blanchard-Fields & Robinson, 1987; Irion & Blanchard-Fields, 1987) found that while older workers exhibited less avoidance of threatening and challenging situations than very young workers, they tended toward escape-type coping behaviors more than middle age workers.

Despite the many difficulties that may hinder the desire to learn new skills among older employees, their willingness to maintain a career built over the years could be a motivator to learn new skills. In a comprehensive review of the age in the workplace literature, Doering et al. (1983) reported a positive correlation between age and internal work motivation. Although the results could be attributed to cohort effects (e.g., Lawrence, 1988), they could also be explained by the attitudes unique to the career stage of the older workers: the maintenance stage. In the maintenance stage, workers are interested in continuing the career established earlier or recently adopted at mid-life (Greller & Nee, 1989). Internal work motivation can be directed toward learning new skills. Fossum et al. (1986, p. 366) summed up the salience of motivating factors through the use of expectancy theory. They argued that "individuals will be 'motivated' to acquire KSAs if: (a) they believe they can acquire them, (b) they perceive that acquiring KSAs will lead to certain outcomes, and (c) they value these outcomes." The motivation component needed to learn new requisite skills can have a tremendous effect on the older worker's retraining success by affecting the decisions to retrain and to persevere training.

Organizational Influences on Learning

New skills training may also be dependent upon other age-related issues. Gist et al. (1988) noted that even though post-training performance of computer skills was significantly lower for older workers, training for computer skills might have been more successful for the older participants if the training had been slower paced (cf. Doering et al., 1983). When making adjustments in training programs to accommodate older employees, it must be acknowledged that not all employees experience the same declines in KSAs as they age. In fact, the degree of variability in skills and abilities increases with age (i.e., there is significantly more variance in individual characteristics for older people than for younger individuals [Schaie, 1988]). Age-related declines in physical and mental abilities happen at different times and in different amounts for different people. Nevertheless, seven components of training have been identified to enhance the learning process of older workers: motivation, structure, familiarity, organization, time, active participation, and learning strategies (Sterns, 1986; Sterns & Doverspike, 1988, 1989). Older workers tend to do best in well structured, self-paced training programs that build on familiar material by gradually increasing complexity. Additionally, intervals of active practice, the integration of learning strategies within the training program, and emphasis on the relevance of the material being taught increase learning (cf. Belbin, 1970; Belbin & Belbin, 1972; Belbin & Downs, 1964, 1966; Downs & Roberts, 1977).

Progress has been made in improving the effectiveness of training programs for older individuals (Sterns & Doverspike, 1989), but whether organizations are willing to adapt training to older workers becomes an overriding issue affecting the learning of new skills necessitated by technological changes in jobs. Organizations may not be providing optimal presentation of prerequisite and requisite information for the older worker to learn new skills. This may be caused by allowing stereotypes to guide perceptions of all older employees. Such an approach is erroneous because aging is a very individual process, and assessment of each individual rather than aggregate assumptions of abilities is necessary.

Another problem is that organizations and managers may be hindering the training of older workers for redesigned jobs. Although blatant discrimination appears to be rare (cf. Miller, Kaspin, & Schuster, 1990), more subtle and even unconscious discrimination appears to exist (Brousseau, 1981). Opportunities for training and feedback needed to correct performance may be withheld from older workers by those harboring conscious or unconscious biases against older workers (Brousseau, 1981; Dalton & Thompson, 1971; Fossum et al., 1986). Both skills training and self-efficacy (and consequently, motivation) may be affected by these practices.

In dealing with the above organizationally induced constraints placed on older employees occupying redesigned jobs, training programs could be altered slightly to accommodate for longer learning curves needed by some older workers. As Gist et al. (1988) noted, a simple alteration such as increasing the length of training programs may enable older workers to perform as well after training as younger workers.

Organizations may also be hindering success of retraining by not addressing attitudes toward the change. Attitudes can be instrumental in affecting behavior (Fishbein & Ajzen, 1975). Older employees tend to respond better to job redesign when changes are introduced gradually (cf. Doering et al., 1983; Gist et al., 1988). The literature on job change has focused on the scope of changes, but has paid little attention to the rate of change (Fossum et al., 1986). Although all employees should respond best to job redesign when they have been given advanced preparation and training, this is especially true of older employees. For example, in a study of auto workers who were gradually introduced to an upcoming technological change, older workers showed more positive attitudes toward the resulting job changes than did younger employees (Herold & Yoon, 1988). Acknowledging the importance of attitudes addresses the motivation and self-efficacy components necessary for new skills learning as well as providing prerequisite and new skills training for the job redesign. In the Herold and Yoon study (1988), employees were exposed to the technological change 18 months before actual implementation on the job. Training was provided across this 18-month period to gradually prepare employees for the upcoming change. This result combined with Gist et al.'s (1988) comment that older employees may have responded better with a slower paced training program, suggests that a key issue may not be the degree of technological change but rather the speed at which it takes place. Even though an organization may be compelled to adopt technologically drastic and complex changes (Greller & Nee, 1989), the organization plays a vital role in preparing adaptation to the changes through the advance dissemination of information concerning the upcoming change and the careful design and timing of training.

Finally, with the abundance of research available to guide the training of older adults, the organization's or manager's decision becomes one of whether training programs tailored to older workers are worth the extended time, expense, and effort involved in implementing them. As mentioned earlier, the attitudes (Herold & Yoon, 1988; Kacmar & Ferris, 1989) and job commitment (e.g., Doering et al.,1983; Greller & Nee, 1989) of older workers are often higher than those of younger workers. Actual levels of performance are discussed in the next section.

In review, the factors affecting learning of new skills are the individual's existing KSAs, motivation, and training circumstances. Different degrees of job redesign result in varying amounts of employee skills obsolescence. Each individual has existing KSAs which may or may not be declining. The individual may or may not be motivated to pursue and persevere training. Likewise, the organization may or may not be willing to provide training for each employee and when provided, the type of training may not be adequate for each individual. The speed, complexity and type of technological changes imposed on the older worker can lead to employee obsolescence if proper retraining for that individual is not offered (London & Bassman, 1989).

ACTUAL EMPLOYEE JOB PERFORMANCE

A question that has led to much empirical research in the gerontology, physiological psychology, and neurological psychology literatures is the extent to which aging causes declines in physical and mental capabilities. Many of the same issues that were addressed in examining capabilities for learning new skills also apply to job performance. Age is negatively related to many physical abilities such as reaction time, vision, and hearing (Droege et al., 1963). However, the relation between age and mental abilities is less clear as "aging does not affect all realms of cognitive ability in a uniform manner" (Ellis & Oscar-Berman, 1989, p. 133). For example, verbal skills often do not decline, but nonverbal communication skills appear to be more susceptible to aging effects (Riege, Metter, & Williams, 1980). Still, much of the decline that does occur in mental and physical abilities occurs at advanced ages (Ellis & Oscar-Berman, 1989), after most individuals have retired from the work force. Evidence indicates that for success on most jobs, the necessary physical and mental attributes do not decline appreciably until individuals have passed the normal retirement age (Brousseau, 1981).

Similarly, research by Perlmutter, Kaplan, and Nyquist (1990) showed age-related deficits on some measures of physical and cognitive abilities, but suggesting age-related increases in job knowledge, that older employees may be "more competent at self-monitoring, social role-taking, and social intelligence in general." In summary, there is no evidence for the supposition that job performance generally declines with age, and for some jobs, performance may actually improve with age (cf. Waldman & Avolio, 1986).

SUPERVISOR PERFORMANCE RATINGS

Regardless of an older employee's actual job performance, it is rated performance that has an influence on personnel decisions regarding that employee. Raters, typically immediate superiors, may allow biases to enter into their ratings (Landy & Farr, 1980). Stereotypes or implicit theories may cause raters to make rating errors in either a positive or negative direction (Ilgen & Feldman, 1983). With respect to older employees, there may be leniency bias when it is assumed that an older employee's experience should translate into higher performance. Similarly, there may be severity bias when the rater assumes that older workers' physical and/or mental capabilities have declined. With respect to technological change, raters may expect older employees to respond less favorably to technological change and thus underrate their job performance. Underrated performance may cause adverse impact against older employees in terms of personnel decisions such as promotion, pay raises, retention, and opportunities for training.

Bias may not only affect performance appraisals, but also have an impact on supervisory behaviors directed at an older subordinate (Feldman, 1986). For example, a supervisor who does not expect an older subordinate to perform well on a job may delegate more routine tasks to this employee and withhold such important resources as job training, feedback, attention, and support. Such resources are especially critical for employees facing technologically changed jobs. A supervisor who holds an implicit theory that older employees are not capable of being retrained for a job that has been redesigned may prevent a subordinate from receiving necessary training. Supervisory behavior that unfairly discriminates against older subordinates may cause declines in actual job performance (cf., Darley & Fazio, 1980; Eden & Ravid, 1982). Difficulties will be exacerbated for older employees whose performance ratings are unfairly biased on top of receiving fewer resources and less support from the supervisor.

Reviews of research examining the correlation between age and performance ratings have shown great inconsistency across studies (McEvoy & Cascio, 1989; Rhodes, 1983; Waldman & Avolio, 1986). The most ambitious of these reviews was McEvoy and Cascio's (1989) meta-analysis of 96 samples obtained by scanning 46 journals over a 22-year period. Results showed an average correlation of .06 with a range from −.44 to +.66. In an attempt to explain this wide range of correlations, several situational variables have been examined as potential moderators of the age-performance rating relation.

Waldman and Avolio (1986) found type of performance measure and job type to have a significant moderating effect on the relation between age and performance ratings. Correlations between age and performance were positive for objective performance measures, but negative for supervisory ratings suggesting the possibility of unfair discrimination. With respect to job type, the correlation between age and performance rating was more positive for professional than for nonprofessional employees. However, with a substantially larger number of studies and total sample size, McEvoy and Cascio (1989) found no support for type of rating or for job type as moderators of the relation between age and performance rating.

One moderator that has been shown to significantly moderate the correlation between age and performance ratings is employee job experience. A meta-analytic study involving 24,219 employees from 111 samples representing many different occupations examined the moderating effect of experience on age-performance rating correlations (Avolio, Waldman, & McDaniel, 1990). These researchers found that partialling experience from the relation between age and performance resulted in nonsignificant partial correlations, but experience-performance correlations removing the effect of age remained significant. Because experience typically correlates highly with age, its effect can become redundant, thus leading to the need to search for other moderaters of the age-performance relation.

In addition to examining overall experience, often defined as job tenure, specific types of previous experience may also be useful to consider when examining age-performance relations. For example, Parsons, Liden, O'Connor, and Nagao (1990) investigated employee reactions to the implementation of computer technology in an insurance industry trade association. They found that although the correlation between age and self-rated performance was negative, the partial correlation removing the effect of prior computer experience was not significant. Extrapolating this finding to other types of technological changes, it is likely that older employees who have faced numerous technologically created job changes in the past will adjust to subsequent technological change with much greater ease than will older employees whose jobs have remained relatively stable in the past (cf. Pinder & Schroeder, 1987).

The potential moderator that has been explored in the current paper is the degree and speed of job design changes resulting from the implementation of new technology. It may be that the correlation between age and performance ratings tends to be negative when there is dramatic and rapid change made in employees' jobs, coupled with no training or inappropriate training. The same correlation may be positive in cases where virtually no change has occurred due to the positive effect that experience has on job performance (e.g., Schmidt, Hunter, Outerbridge, & Goff, 1988) or when the change has been introduced gradually with appropriately paced training. As many authors have pointed out, longitudinal research is needed for proper examination of the relation between age and performance ratings (e.g., McEvoy and Cascio, 1989; Robinson et al., 1985; Sterns & Alexander, 1987). This is especially true for tests of hypotheses concerning the moderating effect of different types and speeds of technologically-induced job changes.

SUMMARY

A model of the process by which technological change influences the performance and rated performance of employees, with emphasis on older employees, has been presented. It was proposed that technological change will typically result in job redesign which in turn causes changes in KSA requirements for affected jobs. It was proposed that older employees do not respond as well to increased job complexity as younger employees unless the change is introduced gradually with appropriately paced training programs. Thus, degree and speed of technological change along with type of job training may significantly moderate the perplexing wide range of correlations between age and job performance. It was stressed that the use of training to provide employees with opportunities to develop new KSAs is typically cost effective and offers organizations an opportunity to prevent obsolescence in an era of predicted labor shortages. Empirical research, especially longitudinal research, is needed to examine the relative importance and possible interactive effects of the variables presented in the proposed model.

REFERENCES

Avolio, B.J., & Waldman, D.A. (1990). An examination of age and cognitive test performance across job complexity and occupational types. *Journal of Applied Psychology, 75,* 43-50.

Avolio, B.J., Waldman, D.A., & McDaniel, M.A. (1990). Age and work performance in non-management jobs: The effects of experience and occupational type. *Academy of Management Journal, 33,* 407-422.

Bandura, A. (1977). *Social learning theory.* Englewood Cliffs, NJ: Prentice-Hall.

Bandura, A. (1982). Self-efficacy mechanism in human agency. *American Psychologist, 37,* 122-147.

Beehr, T.A. (1986). The process of retirement: A review and recommendations for future investigation. *Personnel Psychology, 39,* 31-55.

Belbin, R.M. (1970). The discovery method in training older workers. In H. L. Sheppard (Ed.), *Toward an industrial gerontology* (pp. 56-60). Cambridge, MA: Schenkman.

Belbin, E., & Belbin, R.M. (1972). *Problems in adult retraining.* London, England: Heineman.

Belbin, E., & Downs, S.M. (1964). Activity learning and the older worker. *Ergonomics, 7,* 429-437.

Belbin, E., & Downs, S.M. (1966). Teaching paired associates: The problem of age. *Occupational Psychology, 40,* 67-74.

Blanchard-Fields, F., & Robinson, S.L. (1987). Age difference in the relation between controllability and coping. *Journal of Gerontology, 42,* 497-501.

Botwinick, J. (1977). Intellectual abilities. In J. E. Birren & K. W. Schaie (Eds.), *Handbook of the psychology of aging* (pp. 580-605). New York: Van Nostrand Reinhold.

Bracker, J.S., & Pearson, J.N. (1986). Worker obsolescence: The H.R. dilemma of the 80s. *Personnel Administrator, 31,* 109-116.

Brousseau, K.R. (1981). After age forty: Employment patterns and practices in the United States. In C.L. Cooper & D.P. Torrington (Eds.), *After forty* (pp. 155-208). New York: Wiley.

Campion, M.A. (1988). Interdisciplinary approaches to job design: A constructive replication with extensions. *Journal of Applied Psychology, 73,* 467-481.

Campion, M.A. (1989). Ability requirement implications of job design: An interdisciplinary perspective. *Personnel Psychology, 42,* 1-24.

Campion, M.A., & Thayer, P.W. (1985). Development and field evaluation of an interdisciplinary measure of job design. *Journal of Applied Psychology, 70,* 29-43.

Coates, G.D., & Kirby, R.H. (1982). Organismic factors and individual differences in human performance and productivity. In E.A. Alluisi & E.A. Fleishman (Eds.), *Stress and performance effectiveness* (Vol. 3, pp. 91-140). Hillsdale, NJ: Erlbaum.

Collins, P.D., Hage, J. & Hull, F.M. (1988). A framework for analyzing technical systems in complex organizations. *Research in Sociology of Organizations, 6,* 81-100.

Dalton, G.W., & Thompson, P.H. (1971, September-October). Accelerating obsolescence of older engineers. *Harvard Business Review,* pp. 127-137.

Darley, J. M., & Fazio, R. H. (1980). Expectancy confirmation processes arising in the social interaction sequence. *American Psychologist, 35,* 867-881.

Doering, M., Rhodes, S.R., & Schuster, M. (1983). *The aging worker: Research and recommendations.* Beverly Hills, CA: Sage.

Downs, S.M., & Roberts, A. (1977). The training of underground train guards: A case study with a field experiment. *Journal of Occupational Psychology, 50,* 111-120.

Droege, R.C., Crambert, A.C., & Henkin, J.B. (1963). Relation between GATB aptitude scores and age for adults. *Personnel and Guidance Journal, 42,* 502-508.

Eden, D., & Ravid, G. (1982). Pygmalion vs. self-expectancy: Effects of instructor- and self-expectancy on trainee performance. *Organizational Behavior and Human Performance, 30,* 351-364.

Ellis, R.J., & Oscar-Berman, M. (1989). Alcoholism, aging, and functional cerebral asymmetrics. *Psychological Bulletin, 106,* 128-147.

Feldman, J. M. (1986). A note on the statistical correction of halo error. Journal of Applied *Psychology, 71,* 173-176.

Fishbein, M., & Ajzen, I. (1975). *Belief, attitude, intention, and behavior: An introduction to theory and research.* Reading, MA: Addison-Wesley.

Fossum, J.A., Arvey, R. D., Paradise, C.A., & Robbins, N.E. (1986) Modeling the skills obsolescence process: A psychological/economic integration. *Academy of Management Review, 11,* 362-374.

Gist, M.E. (1987). Self-efficacy: Implications for organizational behavior and human resource management. *Academy of Management Review, 12,* 472-485.

Gist, M.E., Rosen, B., & Schwoerer, C. (1988). The influence of training method and trainee age on the acquisition of computer skills. *Personnel Psychology, 41,* 255-265.

Goldstein, I.L. (1986). *Training in organizations: Needs assessment, development, and evaluation.* Monterey, CA: Brooks/Cole.

Greller, M.M., & Nee, D.M. (1989). *From baby boom to baby bust.* New York: Addison-Wesley.

Griffin, R.W. (1981). A longitudinal investigation of task characteristics relationships. *Academy of Management Journal, 24,* 99-113.

Hackman, J.R., & Oldham, G. (1976). Motivation through the design of work: Test of a theory. *Organizational Behavior and Human Performance, 16,* 250-279.

Hackman, J.R., Pearce, J.L., & Wolfe, J.C. (1978). Effects of changes in job characteristics on work attitudes and behaviors: A naturally occurring quasi-experiment. *Organizational Behavior and Human Performance, 21,* 289-304.

Herold, D.M., & Yoon, B.S.(1988). *The nature of employee beliefs about future automation.* Paper presented at the annual Academy of Management meeting, Anaheim, California.

Ilgen, D.R., & Feldman, J.M. (1983). Performance appraisal: A process focus. *Research in Organizational Behavior, 5,* 141-197.

Irion, J.C., & Blanchard-Fields, F. (1987). A cross-sectional comparison of adaptive coping in adulthood. *Journal of Gerontology, 42,* 502-504.

Kacmar, K. M., & Ferris, G. R. (1989). Theoretical and methodological considerations in the age-job satisfaction relationship. *Journal of Applied Psychology, 74,* 201-207.

Kimmel, D.C. (1988). Ageism, psychology, and public policy. *American Psychologist, 43,* 175-178.

Landy, F., & Farr, J. (1980). Performance rating. *Psychological Bulletin, 87,* 72-107.

Lawrence, B.S. (1988). New wrinkles in the theory of age: Demography, norms, and performance ratings. *Academy of Management Journal, 31,* 309-337.

Loher, B.T., Noe, R.A., Moeller, N.L., & Fitzgerald, M.P. (1985). A meta-analysis of the relation of job characteristics to job satisfaction. *Journal of Applied Psychology, 70,* 280-289.

London, M., & Bassman, E. (1989). Retraining midcareer workers for the future workplace. In I.L. Goldstein and Associates (Eds.), *Training and development in organizations* (pp. 333-375). San Francisco: Jossey-Bass.

Miller, C. S., Kaspin, J. A., & Schuster, M. H. (1990). The impact of performance appraisal methods on Age Discrimination in Employment Act cases. *Personnel Psychology, 43,* 555-578.

McEvoy, G.M., & Cascio, W.F. (1989). Cumulative evidence of the relationship between employee age and job performance. *Journal of Applied Psychology, 74,* 11-17.

Parsons, C.K., Liden, R.C., O'Connor, E.J., & Nagao, D.H. (1990). *Employee responses to technologically driven change: The implementation of office automation in a service organization.* Unpublished manuscript.

Perlmutter, M., Kaplan, M., & Nyquist, L. (1990). Development of adaptive competence in adulthood. *Human Development, 33,* 185-197.

Pierce, J.L. (1984). Job design and technology: A sociotechnical systems perspective. *Journal of Occupational Behaviour, 5,* 147-154.

Pinder, C. C., & Schroeder, K. G. (1987). Time to proficiency following job transfers. *Academy of Management Journal, 30,* 336-353.

Rhodes, S.R. (1983). Age related differences in work attitudes and behavior: A review and conceptual analysis. *Psychological Bulletin, 93,* 328-367.

Riege, W.H., Metter, E.J., & Williams, M.V. (1980). Age and hemispheric asymmetry in nonverbal tactual memory. *Neuropsychologia, 18,* 707-710.

Robinson, P. K., Coberly, S., & Paul, C. E. (1985). Work and retirement. In R. H. Binstock & E. Shanas (Eds.), *Handbook of aging and the social sciences* (pp. 503-527). New York: Van Nostrand Reinhold.

Rousseau, D.M. (1977). Technological differences in job characteristics, employee satisfaction and motivation: A synthesis of job design research and sociological systems theory. *Organizational Behavior and Human Performance, 19,* 18-42.

Roznowski, M., & Hulin, C.L. (1985). Influences of functional specialty and job technology on employees' perceptual and affective responses to their jobs. *Organizational Behavior and Human Decision Processes, 36,* 186-208.

Rumberger, R.W. (1981). The changing skill requirements of job in the U.S. economy. *Industrial and Labor Relations Review, 34,* 578-590.

Salthouse, T.A. (1984). Effects of age and skill in typing. *Journal of Experimental Psychology: General, 113,* 345-371.

Salthouse, T.A. (1988). Effects of aging on verbal abilities: Examination of the psychometric literature. In L.L. Light & D.M. Burke (Eds.), *Language memory, and aging* (pp. 17-35). New York: Cambridge University Press.

Schaie, K.W. (1988). Ageism in psychological research. *American Psychologist, 43,* 179-183.

Schaie, K.W. (1990). Intellectual development in adulthood. In J. E. Dirren & K. W. Schaie (Eds.), *Handbook of the psychology of aging* (3rd ed., pp. 291-309). New York: Academic Press.

Schmidt, F.L., Hunter, J.E., Outerbridge, A.N., & Goff, S. (1988). Joint relation of experience and ability with job performance: Test of three hypotheses. *Journal of Applied Psychology, 73,* 46-57.

Sims, H.P., Szilagyi, S.D., & Keller, R.T. (1976). The measurement of job characteristics. *Academy of Management Journal, 19,* 195-212.

Smith, A.D., & Fullerton, A. M. (1981). Age difference in episodic and semantic memory: Implications for language and cognitions. In D.S. Beesley & G.A. Davies (Eds.), *Aging and communication* (pp. 139-155). New York: Grune & Stratton.

Smith, M. (1981). Over forties in manual and technical work. In C.L. Cooper & D.P. Torrington (Eds.), *After forty* (pp. 59-77). New York: Wiley.

Sparrow, P. R., & Davies, D. R. (1988). Effects of age, tenure, training, and job complexity on technical performance. *Psychology and Aging, 3,* 307-314.

Sterns, H.L. (1986). Training and retraining adult and older adult workers. In J.E. Birren, P.K. Robinson, & J.E. Livingston (Eds.), *Age, health, and employment* (pp. 93-113). Englewood Cliffs, NJ: Prentice-Hall.

Sterns, H.L. & Alexander, R.A. (1987). Industrial gerontology: The aging individual and work. *Annual Review of Gerontology and Geriatrics, 7,* 243-264.

Sterns, H.L., & Doverspike, D. (1988). Training and development of the older worker: Implications for human resource management. In H. Dennis (Ed.), *Fourteen steps to managing an aging workforce* (pp. 97-110). New York: Lexington.

Sterns, H.L., & Doverspike, D. (1989). Aging and the training and learning process. In I.L. Goldstein and Associates (Eds.), *Training and development in organizations* (pp. 299-332). San Francisco: Jossey-Bass.

Stone, E.F., & Gueutal, H.G. (1985). An empirical derivation of the dimensions along which characteristics of jobs are perceived. *Academy of Management Journal, 28,* 376-396.

Taber, T.D., Beehr, T.A., & Walsh, J.T. (1985). Relationships between job evaluation ratings and self-ratings of job characteristics. *Organizational Behavior and Human Decision Processes, 35,* 27-45.

Thompson, J.D. (1967). *Organizations in action.* New York: McGraw-Hill.

Waldman, D.A., & Avolio, B. (1986). A meta analysis of age differences in job performance. *Journal of Applied Psychology, 71,* 33-38.

Walsh, D.A. (1975). Age differences in learning and memory. In D.S. Woodruff & J.E. Birren (Eds.), *Aging: Scientific perspectives and social issues* (pp. 125-151). New York: Van Nostrand.

Weg, R.B. (1975). Changing physiology in aging: Normal and pathological. In D.S. Woodruff & J.E. Birren (Eds.), *Aging: Scientific perspectives and social issues* (pp. 229-256). New York: Van Nostrand.

Zaccaro, S.J., & Stone, E.F. (1988). Incremental validity of an empirically based measure of job characteristics. *Journal of Applied Psychology, 73,* 245-252.

Zenger, T.R., & Lawrence, B.S. (1989). Organizational demography: The differential effects of age and tenure distributions on technical communication. *Academy of Management Journal, 32,* 353-376.

BIOGRAPHICAL SKETCHES OF THE CONTRIBUTORS

Susan M. Adams (MEd, Georgia State University) is a Ph.D. student at Georgia Institute of Technology in the field of Organizational Behavior. Her current research interests include the feedback sending process, training and development issues, and statistical applications to behavioral research. She co-authored and presented papers at the Second International Conference on Managing the High Technology Firm concerning aging workers and technological changes in the workplace and at the 1990 Annual meeting of the Academy of Management concerning performance feedback seeking.

Thomas J. Allen, Jr. is the Gordon Y. Billard Professor of Management at the Sloan School of Mangement, M.I.T. For over 25 years, Dr. Allen has been involved in a series of studies of problem solving and communication behavior of RD&E engineers and scientists. He is most well known for his work on the role of the technical gatekeeper in technology transfer and the influence of architectural layout on human communication behavior. He is currently engaged with Professor Katz in extensive studies of matrix relationships, promotional ladders and reward systems, and the aging of technical groups. He received his Masters in E.E. and his Ph.D. in Management from M.I.T. He is the author of the well-known book *Managing the Flow of Technology*.

Donde P. Ashmos (Ph.D. University of Texas at Austin) is an assistant professor in the division of Management and Marketing at the University of

Texas at San Antonio. Her research interests include behavioral decision making, strategic management processes, and health care management.

Robert S. Atkin joined the Katz Graduate School of Business University of Pittsburgh, in 1987 as a visiting faculty member in the Human Resource Management. Prior to joining KGSB, Atkin was a faculty member at Carnegie-Mellon's Graduate School of Industrial Administration from 1975-1986, serving as its Associate Dean and Director of Masters' Programs for four years (1980-84) Atkin is the author of more than 20 published papers and the co-editor of *Absenteeism*, about which *Personnel Psychology* has said, "Serious students . . . should view this book as mandatory." In addition, he has been Principle or Co-Principal Investigator of 7 funded research contracts producing in excess of $1.3 million. As an instructor, Professor Atkin has taught at various levels (Executive, PhD, MBA, and undergraduate) in areas ranging from Organizational Behavior and Human Resource Management to Statistical Quality Control, Manufacturing Management, and Management Strategy. Four students have earned their PhD degree under his direction. On two occasions, he has been the proud recipient of Teaching Awards. His current research focuses on three themes: (1) the use of Political Action Committees (PACs) by business and labor; (2) the political and corporate implications of the changing demographics of the US workforce; and (3) the changing roles of women in the Japanese workforce. He has also had a long-standing interest in the use of Human Resource policy to establish firm-level comparative strategic advantage.

Joseph C. Bentley, Ph.D. is a professor of management and organizational behavior and adjunct professor of educational psychology at the University of Utah. His areas of research and consulting include organizational change, the communication process in organization, the internalization of values and the importance of meaning and purpose in organizational life. He has published over 40 articles and five books. Currently he is directing a major cultural change program for AT&T.

William S. Brown is a member of the undergraduate faculty of the University of Pittsburgh and the graduate faculty of LaRoche College in Pittsburgh, Pennsylvania. Formerly he was Director of Personnel at Barnard College, Columbia University and Fairleigh Dickinson University. Prior to entering academia he held corporate headquarters human resources positions at the Prudential Insurance Company of America, Philip Morris, Inc., and the Mutual of New York (MONY). He has published over 20 articles, book chapters, and essays in various magazines and professional journals. His research interests include change management, ethics, privacy issues in the workplace and application of management theory in the public sector.

Stephen J. Carroll (Ph.D. University of Minnesota) is a professor of management and organization at the College of Business and Management, University of Maryland. He is also research director of the Dingman Center for Entrepreneurship and is also associated with the Center for Innovation. He is the author and coauthor of numerous well-known books and articles. He is a Fellow of the Academy of Management, a Fellow of The American Psychological Association, and was designated a Distinguished Scholar-Teacher of the University of Maryland.

Susan G. Cohen is a Research Scientist at the Center for Effective Organizations, Graduate School of Business, University of Southern California. She received her B.A. in Psychology from the State University of New York at Buffalo, her M.A. in Applied Behavioral Science from Whitworth College and her M. Phil and Ph.D. in Organizational Behavior from Yale University. Dr. Cohen has done research and consulted on a variety of approaches to improving organizational effectiveness, including group empowerment and effectiveness, employee involvement, organization development and change, participative management, and strategies for executive education and management development. She is particularly interested in team effectiveness in white-collar and service industries. She is the author of several chapters dealing with customer service and top management teams in *Group That Work,* edited by R. Hackman (Jossey-Bass, January 1990).

Garth Coombs, Jr. is a doctoral candidate in Organizational Behavior and Organization Theory at the University of Colorado, College of Business and Administration. Mr. Coombs holds the B.S. degree in Engineering from Stanford University and the M.B.A. from the University of Denver. His research interests center on high technology and on organizational demography.

Amos Drory earned his Ph.D in I/O Psychology from Temple University, Philadelphia. He is currently the chairman of the Department of Industrial Engineering and Management at Ben-Gurion University, Israel. His major research interests are in Organizational Politics, Motivation, and Performance Appraisal.

Dennis Duchon (Ph.D., University of Houston) is an associate professor in the Division of Management and Marketing at the University of Texas at San Antonio. His research has been devoted to behavioral decision making, R&D management, and motivational processes.

Kenneth J. Dunegan (Ph.D. University of Cincinnati) is an assistant professor in the department of Management and Labor Relations at Cleveland State University. His research interests include behavioral decision making, leadership processes, and social processes in the workplace.

Robert H. Faley is Associate Professor of Administrative Sciences in the College of Business Administration at Kent State University. Dr. Faley received his Ph.D from the University of Tennessee and was an Assistant Professor at the Krannert Graduate School of Management at Purdue University. His areas of interest include the human resource policy implications of federal regulation as well as human resource planning, selection, and appraisal. He is coauthor with Rich Arvey of *Fairness in Selecting Employees.*

Anil K. Gupta (Ph.D. Harvard University) is a professor of management and organization at the College of Business and Management University of Maryland. Dr. Gupta previously served on the faculty of Boston University. He is the author of numerous papers in prestigious management journal and is nationally known for his research on business unit strategies within diversified firms. A 1986 survey indicated that one of his papers was the number one cited paper over the previous five year period in strategic management. He is the recipient of the 1991 Glueck Best Paper Award in Business Policy and Strategy from the American Academy of Management.

Ralph Katz is Professor of Management in both the College of Business Administration, Northeastern University and the Management of Technology Department of M.I.T.'s Sloan School of Management. Over the past 17 years, Professor Katz has been carrying out extensive management research on technology-based innovation with a particular interest in the management of technical professionals and project teams. In 1981, the Organizational Behavior Division of the Academy of Management awarded him the "New Concept Award." Dr. Katz received his B.S. from Carnegie-Mellon and his MBA and Ph.D. from the University of Pennsylvania's Wharton School. His most recent book is titled *Managing Professionals in Innovative Organizations: A Collection of Readings.*

Lawrence S. Kleiman is Professor of Management in the College of Business Administration at the University of Tennessee at Chattanooga. He received his Ph.D. in Industrial/Organizational Psychology from the University of Tennessee. Dr. Kleiman's areas of interest include the human resource policy implications of federal regulation as well as the selection, appraisal, and training and development of human resources.

Robert C. Liden is an Associate Professor of Organizational Behavior and of Human Resources Management in the College of Management at Georgia Institute of Technology. After receiving his Ph.D. in organizational behavior at the University of Cincinnati, he conducted a year of postdoctoral research in the Psychology Department at the University of Washington, Seattle. His research on leader-subordinate relations, employment interviews, employee responses to technological change, and aging processes has been published in the *Academy of Management Journal, Academy of Management Review, Journal of Applied Psychology, Journal of High Technology Management, Organizational Behavior and Human Performance,* and *Personnel Psychology.*

Kathryn Martell (Ph.D. University of Maryland) is an assistant professor of management at the School of Business, Southern Illinois University at Edwardsville. She is the coauthor of several published papers on strategic human resources management. Prior to teaching, she was Director of Congressional Affairs for the U.S. Department of Energy and a consultant on strategic issues for domestic and international public and private sector organizations.

Marick F. Masters is currently an Associate Professor of Business Administration at the Joseph M. Katz Graduate School of Business, University of Pittsburgh. He is also Director of MBA Programs at the Katz Graduate School of Business and Coordinator of the Human Resources Management Faculty. Prior to joining the University of Pittsburgh, Dr. Masters was on the faculty of the Department of Management at Texas A&M University. He received his Ph.D. in Labor and Industrial Relations from the University of Illinois in 1983, where he was a University Fellow. Dr. Masters' main teaching and research interests are in the areas of the public-policy regulation of human resource management, human resource management, industrial relations, and corporate and labor political activity. He has taught courses in collective bargaining, arbitration, principles of management, public policies in human resource management, statistics, and human resource management. He has published articles in such journals as the *Academy of Management Journal, Industrial Relations, Industrial and Labor Relations Review, Journal of Labor Research, American Journal of Political Science, Journal of Politics, Public Administration Review, Journal of Collective Negotiations in the Public Sector, Advances in Industrial and Labor Relations Review, Personnel Administrator,* and *Labor Studies Journal.* He has served on the editorial boards of the *Journal of Management* and the *Journal of Managerial Issues.* Dr. Masters has made numerous presentations to a number of professional and business associations, including the Academy of Management, Industrial

Relations Research Association, American Political Science Association, Southwest Academy of Management, Pittsburgh Personnel Association, Society of Federal Relations Professionals, and the Tri-State Employers' Association. He is also an active participant in many of these organizations. Dr. Masters has consulted for various organizations and taught in many executive programs.

Monty Mohrman is the Associate Director of Research and a Research Scientist in the Center for Effective Organizations (CEO). The CEO is in the Graduate School of Business Administration at the University of Southern California and is dedicated to original research concerning the effective integration of organizations and their human resources. Dr. Mohrman's major current interests are: the design of effective organizational systems for human resource management, high involvement management approaches, information technologies in organizations, and the management of team work. Of particular interest to Monty are the ways these elements relate to one another and the larger organizational context. He is currently involved in a number of action research projects directly investigating these issues in organizations. Monty Mohrman is also interested in organization design, organizational change, and social action models for understanding organization behavior. He has published in the areas of organization development, participation in decision making, performance management, quality of work life, and organizational impacts of information technology. He is the first author of two books, *Designing Performance Appraisal Systems* and *Large-Scale Organizational Change,* and he is a co-editor of *Doing Research that is Useful For Theory & Practice.* Prior to his present position, Dr. Mohrman was a faculty member in the College of Administrative Sciences at the Ohio State University. He earned his Ph.D. in Organizational Behavior from the Graduate School of Management at Northwestern University. His undergraduate degree is from Stanford University in Physics. Monty has extensive research and consulting experience in both public and private and private sector organizations.

Susan A. Mohrman is a Senior Research Scientist at the Center for Effective Organizations in the Graduate School of Business at the University of Southern California. She received her A.B. in psychology from Stanford University in 1967 and her Ph.D. in Organizational Behavior from Northwestern University in 1978. She has published papers in professional journals and books on the topics of: (1) employee involvement, (2) innovative approaches to the design of organizations, (3) organization development and change, (4) high technology organization, (5) union management cooperative projects, and (6) innovative research and evaluation methodologies. She is an editor of *Research for Theory* *and Practice* (Jossey-Bass, 1985), *Large Scale Organizational Change* (Jossey-Bass, 1989), and *Managing Complexity in High Technology Organizations* (Oxford University Press, 1989). She is co-author of *Self-Designing Organizations: Learning How to Create High Performance* (Addison-Wesley, 1989) and of *Employee Involvement in America: A Study of Contemporary Practice* (American Productivity and Quality Center, 1989).

Robert A. Page, Jr. received his B.S. in Industrial and Labor Relations from Cornell University and his Master of Organizational Behavior from Brigham Young University. He is currently a Ph.D. candidate at the University of California, Irvine. His reserch interests include the management of innovation, institutionalization of organizational change, organizational politics and intergroup dynamics.

Joseph G. Rosse is an assistant professor of Strategy and Organization Management at the University of Colorado at Boulder. He received his Ph.D. in Industrial/Organizational Psychology from the University of Illinois in 1983, and now teaches, conducts research and consults in the areas of employee staffing and retention and employee counter-productive behaviors.

Gerald Schoenfeld is a candidate for a Ph.D. degree at the Joseph M. Katz Graduate School of Business at the University of Pittsburgh. He received his B.A. degree in Psychology from Rutgers University in New Jersey and his M.S. degree in Industrial Psychology from the University of Central Florida. Prior to coming to the University of Pittsburgh, Gerald held numerous Human Resource positions with a large, northeastern bank, most recently the position of Salary Supervisor, Bank Officer. In addition, Gerald has consulted extensively with many Fortune 500 companies in the areas of pay for performance, assessment centers, and performance review systems. Currently, he teaches human resource courses at the University of Pittsburgh. Gerald's research interests are in the areas of pay for performance, strategic utilization of human resource assets, the ratee's role in the performance apparaisal process and the use of Political Action Committees. His research has appeared in such journals as *Human Resource Planning* and *Labor Studies Journal.*

Gregory K. Stephens is an assistant professor in the department of organizational behavior at the M.J. Neeley School of Business at Texas Christian University. He received his B.S. and M.B.A. degrees at Brigham Young University, and is a Ph.D. candidate at the University of California, Irvine. He has published in the area of careers in the *Journal of Management* and the *Journal of Management Studies*, and contributed a chapter titled "Women in Management Around the World" for the forthcoming book

Pathways to Excellence: New Patterns for Human Resource Utilization. He is currently involved in research on career transitions, international human resources, and dual-career issues.

George C. Tokesky, the manager of Compensation at Kennametal Inc., is responsible for worldwide compensation management. He has ten years experience in compensation planning and administration, and twenty-five years experience in management. Mr. Tokesky is a Doctoral Candidate, at the Joseph M. Katz Graduate School of Business, University of Pittsburgh, he has received Bachelor's degrees from the University of Pittsburgh (in Psychology) and John Carroll University (in English literature), and an M.B.A. from the University of Pittsburgh.

Angela Tripoli is a doctoral student in organizational behavior at the University of California, Irvine. Her research interests include employee-organization psychological contracts, organizational commitment, work strategies, and organizational socialization and disaffiliation processes.

Michael Tushman is Professor of Management and Director of the Innovation and Entrepreneurship Research Center at the Graduate School of Business, Columbia University. His research and teaching interests include the management of innovation and technology, communication networks and technology transfer, and organizational change. Dr. Tushman received the William H. Newman Award for excellence and innovation in teaching and he was the recipient of the California Management Review's first Pacific Telesis Foundation Award. He received his Ph.D. from the Sloan School of Management at M.I.T. and is the author of many books, including *Readings in the Management of Innovation.*